Orange Wave Digital, LLC: Learn ACT Math: The Workbook (3rd Edition)
©2021, Mandee Boster

All rights reserved. No part of this work may be reproduced or distributed in any form by any means - graphic, electronic, or mechanical, including photocopying, recording, taping, or storing in information storage and retrieval systems - without written permission from the publisher.

If you are a parent or teacher, you may duplicate pages for yourself and students in your immediate household or classroom. Please do not duplicate pages for friends, relatives outside your immediate household, or other teachers' students.

The information contained in this book is for informational purposes only. The information provided in this book is not guaranteed to be accurate, but was prepared to the best of the author's knowledge.

This book does not contain actual American College Testing (ACT) questions nor was it produced by the ACT. The information and questions contained in this book are based on past ACT tests and practice tests.

Users of this book are encouraged to do their own due diligence in preparing and taking any academic exams. By reading this book, you agree that the author and TeachMathWithMe.com are not responsible for the success or failure of your academic testing related to any information presented in this book.

Published by
Orange Wave Digital, LLC
701 Highway 31 South
Suite 119
Hartselle, AL 35640

www.TeachMathWithMe.com

Preface

This book was written to help introduce students to the topics and types of questions they may encounter on the math portion of the ACT. I enjoy working math questions, so helping students learn math concepts is important to me. I want all students that desire to do well on the ACT have the practice to perform their best.

This 3rd edition has been updated in 2021 to include worked out solutions, as well as minor grammar updates.

On my site, TeachMathWithMe.com, you will find the materials that I have developed to help students and instructors in the ACT journey. Check those out for additional help.

I hope the student preparing for the ACT finds this book to be helpful! Please feel free to contact me with questions or comments at contact@teachmathwithme.com. I wish the students (and instructors) all the best!

Mandee Boster

About the Author

Hi everyone! I'm Mandee. If it is not obvious by the subject of this book, I love math! Math is something that has always come easy for me...except for maybe in 3rd grade when I thought I had reached my limit in math with FRACTIONS! I have a vivid memory of thinking it was all over right there. But do you know what? The next year in 4th grade a lightbulb went off, and I thought fractions weren't so bad after all. I just needed a little time to let the concept of fractions sink in.

Thinking back, I realize that this happens quite often in math. Some people get a concept first off, but others, it takes a little time. My number one recommendation when helping students is to practice! That is what this workbook and my other resources at TeachMathWithMe.com are all about.

I have a BS and MS in Industrial and Systems Engineering, and in my first career, I worked for NASA planning payload experiments and testing payload software for the International Space Station. It was exciting! But after having children, our family decided to make a pivot, and I became a stay-at-home mom. Now, I am a work-from-home mom that gets to work math problems and help students each day!

In my free time, I enjoy crafting and spending time with my family. I often wear crimson and white and plan to one day live full-time at the beach!

What's Different In this Edition

Besides a few corrections and re-arranging words here and there, the majority of the updates to this edition is the addition of a solutions manual with worked through questions. I hope this is a very welcome update!

Introduction

The math portion of the ACT contains 60 questions. The student has 60 minutes to complete this section. Your first practice math problem in this book is to determine what your average time per question should be to complete each question on the ACT. If you answered "1 minute", you are correct!

But, averaging 1 minute per question means you finish with no time remaining (no checking time). Also, not all questions are created the same in terms of difficulty. It is a good goal to aim for less than a minute per question. This will give you time to answer each question, then either check back through your answers or work on any questions in which you need longer time to complete.

The test usually begins with easier questions and gets more challenging near the end (around #40). As for your timing, if you answer the easier questions fairly quickly (around 30 seconds or less), then you can build up time to use on the more challenging or multi-step questions that require more time.

Calculators are permitted on this section, but make sure that you check the ACT website for guidelines.

Question Content

There are 8 reporting categories in the math section. The 60 questions are broken into 2 main categories: Integrating Essential Skills (40-43%) and Preparing for Higher Math (57-60%). In addition, greater than 25% of all the questions contain a modeling aspect. This means there are "questions that involve producing, interpreting, understanding, evaluating and improving models" (2016/2017 Preparing for the ACT Test).

In the Integrating Essential Skills category, the questions contain skills that students have been exposed to usually prior to 8th grade. Throughout the test, these skills will be tested at higher levels of complexity, so a basic understanding of them is necessary for success. In the upcoming sections, you will see the types of questions that may appear in this category.

The Preparing for Higher Math category picks up with skills usually learned beginning with Algebra and moving beyond. This category contains 5 subcategories: Number & Quantity, Algebra, Functions, Geometry, and Statistics & Probability. See the pie chart for the usual breakout of the content. These numbers are on average and may be different on your actual test. We will see examples from each subcategory in the upcoming sections.

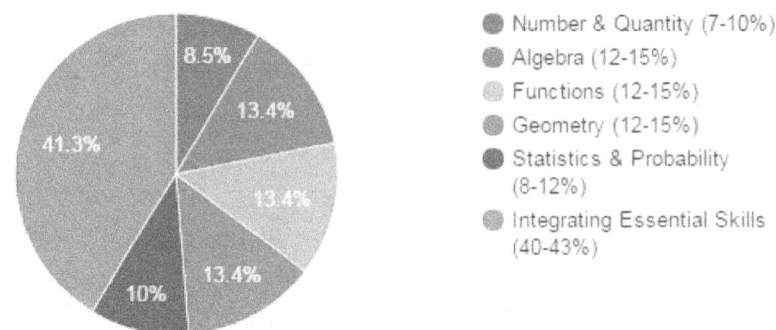

Question Types

The questions on the math section are all multiple choice with 5 options to choose from.

You can expect to see many different types of questions on the math section. There will be word problems, solve/simplify, graphs, figures, and other math problems. The more you practice you will gain confidence and get comfortable working these different types of problems and applying the concepts with different approaches.

Example Questions

The following example questions represent the types of questions that you may find on the ACT math section. Based on the categories described in the previous section (per the "2016/2017 Preparing for the ACT Test" published by ACT.org), we have attempted to categorize our example problems in the same manner. Since there is some overlap in the areas, our categorization may not match how ACT may categorize a question. Overall, this should not make a difference because no matter the categorization, it is the content that needs to be practiced and learned. We have plenty of content to go over in this workbook!

So, let's start going through some examples!

Integrating Essential Skills

The concepts that are found in questions in this category are usually introduced prior to 8th grade. You will notice varying degrees of difficulty in these questions though.

Understanding the basic skill is important, but be prepared to apply the skills to more advanced applications and modeling questions.

Most of these types of questions are considered the "easier" questions. It is important that you grasp these concepts and answer these types of questions quickly, so that you can save time for the more challenging problems. Remember that these questions can be dispersed throughout the entire test, not just at the beginning.

Average/Mean, Median, and Mode

As part of the ACT Mathematics instructions, there is a statement that reads "The word *average* indicates arithmetic mean." So, for the purposes of this test, the terms "average" or "mean" may be interchanged.

1. Matt has the following grades in his science class: 92, 85, 96, and 100. If he has one more test to take in the class, what must he make to finish the grading period with an average of 90?

    A. 77
    B. 79
    C. 82
    D. 90
    E. 93

    > Hint: How many total points does Matt need for a 90 average with 5 test grades?

2. The baseball team sold meal tickets to raise money for the season. If the following number of tickets were sold by each of the players, what is the mean, median, and mode for ticket sales: 20, 15, 32, 22, 25, 20, 24, 27, 20, 18?

    A. 20.3, 21, 20
    B. 25.3, 21, 18
    C. 22.3, 21, 20
    D. 27.1, 21, 20
    E. 27.1, 25.3, 21

    > Average/Mean - sum of the numbers divided by the total number
    > Median - middle number in the set
    > Mode - number that occurs most often in the set

3. Susan is an advertising manager and prepares a monthly report for her supervisor. Susan's monthly sales are in the chart below. What is the average number of Susan's ads per magazine, to the nearest 0.1?

Number of Susan's Ads in a Magazine	Number of Magazines with this Total
1	3
2	5
3	7
4	4
5	2

- A. 2.9
- B. 3.2
- C. 5.1
- D. 7.0
- E. 8.4

Factors

The factors of a number are the numbers that you can multiply together to get the original number. Determining the factors of a number is a basic skill that you could see in advanced applications.

4. What are all the positive factors of 10?

   - A. 1, 10
   - B. 1, 2, 5, 10
   - C. 2, 10
   - D. 10, 20, 30
   - E. 5, 10, 15

5. Which of the following lists all the positive factors of 9?

   - A. 1, 9
   - B. 3
   - C. 3, 6
   - D. 1, 3, 9
   - E. 9, 18, 27

> Review how to use positive and negative numbers in adding, subtracting, multiplying, and dividing.

Least Common Multiple

When you multiply a number by an integer (whole number), the result is a multiple. To find the common multiple between two or more numbers, compare the multiples of each number.

6. What is the least common multiple of 5, 10, and 25?

    A. 2
    B. 10
    C. 25
    D. 50
    E. 100

> In most common multiple questions, you are asked to find the least common multiple. That is the first multiple that the set of numbers have in common.

7. The local deli serves pasta salad every 4 days and pretzel salad every 6 days. If pasta salad and pretzel salad are both on today's menu, how many days before they are both on the menu again?

    A. 1
    B. 4
    C. 6
    D. 8
    E. 12

Fractions

Using fractions is something you have probably done since elementary school. Even though we are introduced to fractions at an early age in our academic careers, they sometimes make math problems appear complicated.

In most cases, a problem is easier to work if we can eliminate the fraction, for example, by multiplying through by a common denominator.

8. Put the following fractions in increasing order: $\frac{1}{2}, \frac{3}{8}, \frac{1}{4}, \frac{5}{16}, \frac{7}{8}$

   A. $\frac{1}{4}, \frac{5}{16}, \frac{3}{8}, \frac{1}{2}, \frac{7}{8}$

   B. $\frac{1}{2}, \frac{5}{16}, \frac{7}{8}, \frac{1}{4}, \frac{3}{8}$

   C. $\frac{7}{8}, \frac{1}{2}, \frac{3}{8}, \frac{5}{16}, \frac{1}{4}$

   D. $\frac{1}{4}, \frac{1}{2}, \frac{3}{8}, \frac{5}{16}, \frac{7}{8}$

   E. $\frac{1}{4}, \frac{1}{2}, \frac{3}{8}, \frac{7}{8}, \frac{5}{16}$

> When preparing for the ACT, determine which is your bigger challenge: knowing the content or working the problems quickly.

9. When 6 ¾ is written as an improper fraction, in lowest terms, what is the numerator?

   A. 3
   B. 18
   C. 24
   D. 27
   E. 30

> Remember that a fraction can be thought of as the part over the whole. The number above the fraction bar is the *numerator*, and the number below the fraction bar is the *denominator*.

10. What is the least common denominator for adding $\frac{1}{6}, \frac{3}{8}, \frac{5}{12}$?

    A. 6
    B. 12
    C. 24
    D. 48
    E. 576

Reciprocals

The *reciprocal* of a number is the inverse of a number. Every number has a reciprocal except for 0 (1/0 is undefined...you cannot have 0 as a denominator).

11. If 2 reciprocals are multiplied together, what is their product?

    A. 0
    B. 1
    C. 2
    D. 5
    E. 10

    > Hint: Pick a few reciprocals and multiply them together. See what you get as an answer.

12. What is the reciprocal of ⅝ ?

    A. $\frac{1}{8}$
    B. $\frac{5}{8}$
    C. $\frac{8}{5}$
    D. 5
    E. 8

    > The reciprocal of a whole number is 1 divided by the whole number.
    >
    > For fractions, swap the numerator and the denominator.

Percent

Percent is a ratio that means parts per hundred. To change a decimal to a percent, you must move the decimal 2 places to the right and add the "%" sign. In turn, to change a percent to a decimal, move the decimal 2 places to the left and take away the "%" sign.

13. If a $350 laptop is discounted 15%, but then has 9% sales tax added, what is the total price you must pay?

    A. $31.50
    B. $297.50
    C. $324.28
    D. $381.50
    E. $402.50

Use the following table for questions 14 and 15.

Menu Item	Number of Favorite Votes
Pizza	19
Hamburger	9
Salad	5
Chicken Tenders	14
Sloppy Joes	3

14. A poll of 50 random students in a high school was taken concerning a favorite lunch menu item. The results are in the table above. What percent voted for chicken tenders?

   A. 14%
   B. 20%
   C. 25%
   D. 28%
   E. 30%

> Your goal should be to finish the math section with time remaining to check.

15. If this poll is indicative of how the 800 students in the entire school would vote, what is the best estimate of the number of votes hamburger would receive from the entire school?

   A. 9
   B. 50
   C. 144
   D. 150
   E. 450

16. As a jewelry consultant, Sue makes a commission off each item she sells. If her sales are $500, she makes a commission of $150. How much commission does she earn if she makes $850 in sales?

    A. $150
    B. $ 255
    C. $ 300
    D. $ 500
    E. $ 850

17. Julie visited a craft store where all items were 15% off. She wants to program her calculator so she can input the marked price and the discounted price will be output. What is the expression for the discounted price on a marked price of *p* dollars?

    A. 0.15p
    B. 0.15p + p
    C. 0.15p + 0.85p
    D. p - 0.15p
    E. 0.15p - p

> It is important to understand that a percent is a portion of a total, in parts per one hundred.

18. 250 students entered an art contest. The entries are divided into 4 categories as shown below. There are 50 prizes to be awarded. If the prizes are to be awarded in proportion to the number of entries in each category, how many prizes should be in the 10-11 year old category?

Age Category	Under 9	10-11	12-13	14-15
Number of Entries	70	80	65	35

    A. 13
    B. 16
    C. 20
    D. 25
    E. 80

Ratios

A *ratio* is a comparison between two or more values.

19. A group of 3 friends share and eat a whole pizza. Carrie eats 1 piece, John eats 4 pieces, and Sam eats 3 pieces. What is the ratio of Carrie's share to John's share to Sam's share?

    A. 3:4:1
    B. 4:3:1
    C. 7:1
    D. 1:8
    E. 1:4:3

> Ratios are usually shown in fraction form or using ":".

20. Jane fills a bowl with colored jelly beans. There are 20 red, 16 yellow, 10 green, and 14 purple. In lowest terms, what is the ratio of red to yellow to green to purple jelly beans?

    A. 1:4:6:2
    B. 10:8:5:7
    C. 8:5:7:10
    D. 5:4:2:3
    E. 10:16:1:14

> You can think of a ratio as showing you how one item compares in size, price, etc. relative to something(s) else.

Rational Numbers

A *rational number* is any number that can be expressed as the quotient or fraction of two integers. The questions you may encounter on the test related to rational numbers can include the instructions to choose a rational number as a solution option to only use rational numbers in the calculations. Knowing the definition of a rational number, along with keeping some examples in mind, should help you in any questions that make use of rational numbers.

21. Which of the choices is NOT a rational number?

    A. $\sqrt{\frac{5}{25}}$

    B. $\frac{1}{3}$

    C. $\frac{1}{2}$

    D. 5

    E. .21212121…

22. What rational number is halfway between $\frac{1}{2}$ and $\frac{1}{3}$?

    A. $\frac{1}{4}$

    B. $\frac{1}{5}$

    C. $\frac{5}{12}$

    D. $\frac{3}{8}$

    E. $\frac{7}{12}$

Scientific Notation

Scientific notation is a method of writing numbers in terms of a number and/or decimal multiplied by a power of 10.

Here's a way that I remember the exponents: A small number, like a decimal, is written in scientific notation with a negative exponent on the power of 10. A larger number will have a positive exponent on the power of 10.

23. What is 0.00000325 in scientific notation?

    A. 3.25
    B. 325
    C. 3.25×10^{-6}
    D. 3.25×10^{6}
    E. 3250

> To convert a number to scientific notation, the standard is to move the decimal between the first two non-zero digits that you encounter reading from left to right. If you move the decimal to the left, the exponent on the power of 10 should be the positive number of decimal moves. If you move the decimal to the right, the exponent on the power of 10 should be the negative number of decimal moves.

24. What is 438000000000 in scientific notation?

    A. 4.38×10^{-11}
    B. 4.38×10^{-4}
    C. 4.38
    D. 438
    E. 4.38×10^{11}

> To convert from scientific notation back to basic number form, move the decimal the number of positions based on the exponent on the power of 10. A negative exponent indicates moves to the left. A positive exponent indicates moves to the right.

Data Interpretation

The ability to read data from a chart or graph and apply that data is a skill that you may need to use on the test. Be sure to look at units or any chart labeling to make sure that you understand what the data is representing.

25. Jenn is planning to buy cookies from a bakery for her cousin's birthday party. She needs 50 cookies. When she gets to the bakery, she sees the chart below. What's the minimum price Jenn will pay for 50 cookies?

Individual	1 Dozen	2 Dozen
$0.50	$5.50	$10.75

    A. $10.75
    B. $21.50
    C. $22.50
    D. $23.00
    E. $25.00

Rates

A *rate* is a ratio where the terms in the ratio have different units. For example, miles/hr is a rate. In most all the practice and real ACT tests that I have evaluated, there are some type of rate questions, usually in the form of a word problem. Therefore, it is a good idea to be familiar with rate problems.

26. Mike is driving to see the New York Yankees play. It is 300 miles to the interstate exit that he needs to reach. He drives an average speed of 65 miles/hour. If he has already driven 3 hours, but wants to get to the exit in 1.5 more hours, how many miles per hour faster should he drive?

    A. 1.5 miles/hr
    B. 5 miles/hr
    C. 50 miles/hr
    D. 65 miles/hr
    E. 70 miles/hr

27. A machine makes 10 widgets per minute. A second machine makes 25 widgets per minute. The second machine starts making widgets 3 minutes after the first machine starts. Both machines stop making widgets 10 minutes after the first machine started. Together, the 2 machines make how many widgets?

    A. 35
    B. 100
    C. 200
    D. 275
    E. 350

Perimeter

The perimeter of a figure is the distance around the figure. The perimeter of a circle is the circumference. For other shapes, we use the term perimeter, and will need to know the lengths of the sides to calculate this distance.

28. Figure ABCD is a square with side length of 4. Point E is the midpoint of AB, point F is the midpoint of BC, and point G is the midpoint of CD. What is the perimeter of the region AEFGD?

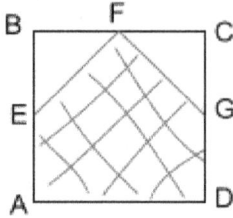

- A. 8
- B. $4 + 2\sqrt{2}$
- C. $8 + 4\sqrt{2}$
- D. 16
- E. $12 + 2\sqrt{2}$

29. The perimeter of a parallelogram is 32 ft. The length of one side is 9 ft. What are the lengths, in ft., of the other 3 sides?

- A. 4, 4, 4
- B. 9, 4, 4
- C. 9, 7, 7
- D. 8, 8, 8
- E. Cannot be determined

> If you feel like you have answered a question 100% correct, note that (maybe by circling the question number in your test booklet). When checking your answers at the end, don't check those circled as "knowing" until you have checked through all other questions first.

30. In the figure below, adjacent sides meet at right angles and lengths given are in feet. What is the perimeter of the figure, in feet?

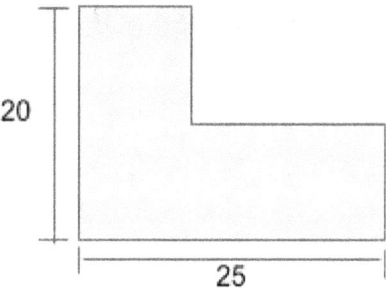

- A. 20
- B. 25
- C. 45
- D. 90
- E. Cannot be determined

Circumference

The circumference of a circle is the distance around the circle. To calculate the circumference, use the following formula:

$$C = 2\pi r = \pi d, \text{ where r is the radius and d is the diameter}$$

31. What is the circumference of a circle with radius of 3 units? Use 3.14 for π.

- A. 4.71
- B. 5.32
- C. 9.42
- D. 18.84
- E. 56.52

> In the questions, be sure to glance at the answer choices to see if π or 3.14 is used in the answer. If you are to substitute in the approximated value, it will usually tell you in the question.

32. An 8.5" diameter bowling ball is released at the foul line and starts rolling towards the head pin. If it is 60 feet from the foul line to the head pin, how many revolutions will the ball make before it hits the head pin?

   A. 12.2
   B. 13.35
   C. 26.69
   D. 26.98
   E. 30

Area - Rectangle

A rectangle is a quadrilateral that is a parallelogram with all four angles equal to 90°. When we are calculating the area of the rectangle, we want to determine how much space is within the rectangle with side length, l, and side width, w. So, the area of the rectangle is l times w. ($A_{rect} = lw$)

33. Mike plans to use grass seed in the backyard of his newly built home. His backyard is a 36 ft x 40 ft rectangular plot, but he has installed a 12 ft x 24 ft inground pool. If each bag of grass seed covers 375 ft², how many bags will he need to purchase?

   A. 2
   B. 3
   C. 4
   D. 5
   E. 6

34. What percent of the area of the larger rectangle is the area of the smaller, shaded rectangle?

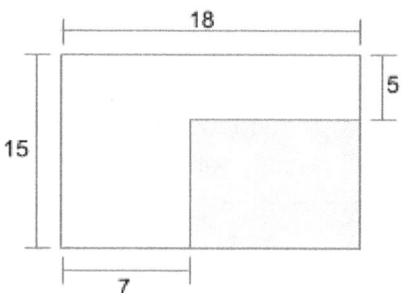

   A. 25.3%
   B. 32.7%
   C. 40.7%
   D. 42.7%
   E. 57.6%

35. A rectangle has an area of 36 ft² and a perimeter of 26 feet. What is the longest of the side lengths, in feet, of the rectangle?

    A. 4
    B. 5.5
    C. 6
    D. 9
    E. 10

    > Hint: Write out the formulas for perimeter and area of a rectangle and, using the unknowns, solve one equation for the other one.

36. In the figure below, the vertices of rectangle ABCD have (x,y) coordinates. What is the area of rectangle ABCD?

    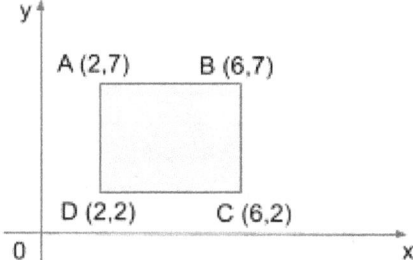

    A. 10
    B. 18
    C. 20
    D. 22
    E. 30

Area - Circle

To calculate the area of a circle, you will need the radius. Sometimes you are directly given the radius, and other times you will have to calculate it from other measures, such as from the diameter or the circumference. The formula for the area of a circle is $A_{circle} = \pi r^2$.

37. What is the area, in m², of a circle with diameter 6 m? Use 3.14 for π.

    A. 9.42
    B. 12.52
    C. 18.84
    D. 28.26
    E. 113.04

38. Jill installed a 9 ft. leash to a stake right outside her front door. What is the outside area, in square ft., the dog can reach from the stake?

   A. 4.5π
   B. 9π
   C. 40.5π
   D. 81π
   E. 105π

Area - Triangle

The area of a triangle is $A_{triangle} = \frac{1}{2}bh$, where b is the base of the triangle and h is the height of the triangle. Obtaining the base and height from a right triangle is very straightforward.

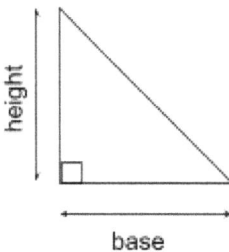

When you do not have a right triangle, you have to "create" a right triangle to determine the height. To do this, draw a straight line from a vertex to form a right angle with the opposite side. That is the height of the triangle.

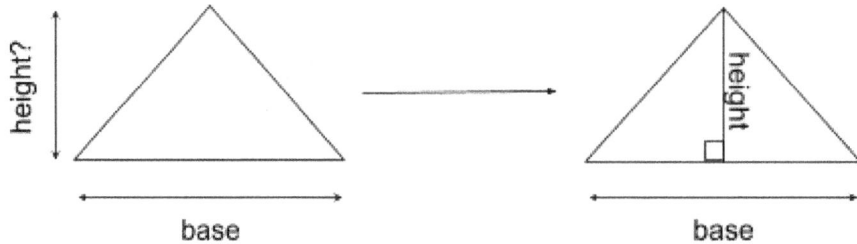

For the examples in this section, making use of the Pythagorean Theorem ($a^2 + b^2 = c^2$) will help to determine the height.

39. What is the area of the right triangle with side lengths 5 cm and 7 cm?

   A. 12.0
   B. 17.5
   C. 19.3
   D. 35.0
   E. 36.2

40. Given the figure below, what is the area of the equilateral triangle ABC?

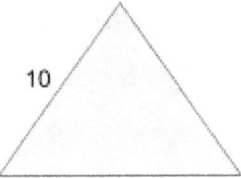

   A. 5
   B. $25\sqrt{3}$
   C. 30
   D. 75
   E. 100

> Be familiar with the terms for a triangle with no sides equal (scalene), only 2 sides equal (isosceles), and all 3 sides equal (equilateral).

41. In the figure below, the vertices of △ABC have (x,y) coordinates. What is the area of △ABC?

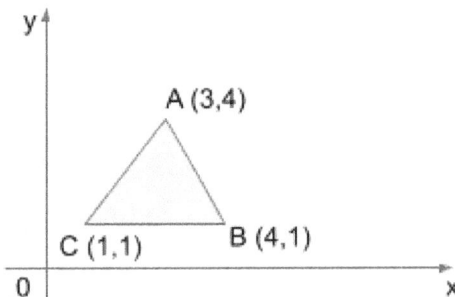

   A. 2
   B. 3
   C. $\dfrac{9}{2}$
   D. 4
   E. Unable to determine

Area - Other

There are many different figures that you may encounter and be asked to determine the area. It is a good idea to try to remember the basic figures' formulas so that you can apply them to other figures. You may have to do this in pieces though. Also, be aware of specifics given on the figures that may make the calculations easier, such as if it is a regular figure or if certain angles or sides are congruent.

42. In the isosceles trapezoid given below, the parallel lines are 8 in and 14 in long, respectively. What is the area of the trapezoid, in in²?

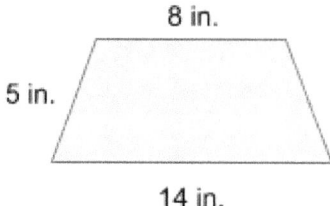

A. 24
B. 32
C. 44
D. 48
E. 52

43. Given the measurements in the figure below and that ∠A = ∠D and BC is parallel to FE, what is the area, in square units, of the hexagon?

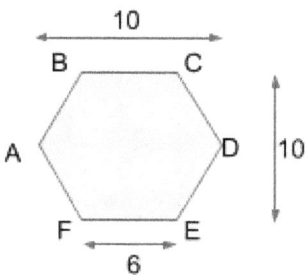

A. 20
B. 60
C. 70
D. 80
E. 90

Midpoint

The midpoint of a line segment is the point halfway between the endpoints of a line segment. It cuts the line segment into 2 equal sections.

44. What is the midpoint of the line with endpoints (1,2) and (7, -2)?

A. (0,1)
B. (1,0)
C. (3, 2)
D. (4,0)
E. (4,1)

> On the coordinate plane, this results in the midpoint being an ordered pair, (x, y).

45. The number line below is the graph of which of the following inequalities?

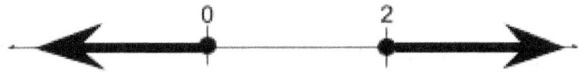

A. $0 \geq x \geq 2$
B. $0 \leq x \leq 2$
C. $x \geq 2$
D. $x \leq 0$
E. $-1 < x < 2$

> Usually instead of getting a single (or even multiple at times) answer when solving the equation, for an inequality, your solution can be in the form of a set, in the form of a range, or as a set of values on a number line.

Preparing for Higher Math

The concepts that are found in questions in this category are ones that students are probably encountering in high school level math classes beginning with Algebra. These questions may also utilize skills from previous years as math builds upon itself.

In the following sections, you will see examples for each of the 5 subcategories, along with those that also test modeling.

These question types may be dispersed throughout the entire test, so be prepared to work them any time within the test period.

Number & Quantity

The Number & Quantity subcategory accounts for between 7 and 10% of the questions on the ACT.

We will work the following types of problems in this section:
- Absolute Value
- Multiplying and Dividing Variables
- Exponents
- Imaginary Numbers
- Matrices
- Complex Numbers

- Vectors

Absolute Value

Absolute value refers to the magnitude of a number without regard to sign. In simplest terms, it is the value of the number no matter if it is a positive or negative number.

46. Evaluate the expression: | (3)(-2) - (6)(4) | - | (2)(-1) + (3)(5) |.

    A. -43
    B. -17
    C. 10
    D. 17
    A. 43

47. -3 | 7 - 5 | + 2 | -3 - 2 | = ?

    A. -72
    B. -16
    C. 1
    D. 4
    E. 7

> When you choose an incorrect answer, is it because you do not know how to work the problem, you used the wrong method to solve the problem, or you made a careless error? Determining this will help you know how to prepare for this type of question for next time.

Multiplying and Dividing Variables

You can only multiply and divide like variables. What is different with multiplying and dividing compared to adding and subtracting with variables is that you can multiply and divide like variables with different exponents. There are a few guidelines to remember though. If you are multiplying same variables, add the exponents. When dividing same variables, subtract the exponents. Also remember that if you have variables in the denominator of a fraction, if you move them up to the numerator, you must multiply the exponents by "-1".

48. What is the product of $(3xy^2)(-2x^3y^3)$?

    A. $-6x^3y^6$
    B. $-6x^4y^5$
    C. $3x^3y^6$
    D. $6x^4y^5$
    E. $6x^3y^6$

TeachMathWithMe.com

49. Evaluate the expression $\dfrac{x^3 y^2 z}{x^{-2} z^3}$.

   A. $\dfrac{x^5 y^2}{z^2}$

   B. $\dfrac{x^6 y^2}{z^2}$

   C. $x^5 y^2$

   D. $\dfrac{x^6 y^2}{z^3}$

   E. $\dfrac{x^5 y^2}{z^3}$

Exponents

When exponents are used, a term or expression is being raised to a power. You can also think of that term or expression being multiplied together the number of times as the exponent. When variables with exponents are raised to a power, then you multiply the exponents to get the new variable exponent value. Also, if you have the same bases, the exponents are equivalent in equations.

50. What is the following expression equivalent to? $\dfrac{(3x^2 y)^3}{xy^2}$

   A. 8xy
   B. $9x^5 y$
   C. $\dfrac{3x^5 y}{xy}$
   D. $27x^5 y$
   E. xy

> If you raise a variable with an exponent to a power, multiply the exponent by the power in which it is raised.

51. What is $(2a^2b)^3$?

    A. $2a^6b^3$
    B. $2a^5b^4$
    C. $4a^5b^4$
    D. $8a^6b^3$
    E. $10a^6b^3$

Imaginary Numbers

The square root of a negative number is an imaginary number, i. An imaginary number is a complex number that can be written as a real number multiplied by i. The imaginary unit, i, is defined by the property that $i^2 = -1$.

52. For $i^2 = -1$, $(2 + i)^2 = ?$

    A. 3
    B. $4 + 4i$
    C. $2 + 3i$
    D. $3 + 4i$
    E. $4i$

> When multiplying or dividing with the imaginary number, i, you can almost treat it as you would a variable, but do not forget the i^2 property.

Matrices

A matrix is an array of numbers. The elements of a matrix are set up in rows and columns.

53. Multiply the following matrices. $\begin{pmatrix} 3 & 2 & 1 \\ 0 & 4 & 2 \end{pmatrix} \begin{pmatrix} 1 \\ 2 \\ 3 \end{pmatrix}$

    A. $\begin{pmatrix} 10 \\ 14 \end{pmatrix}$

    B. $\begin{pmatrix} 7 & 10 \end{pmatrix}$

    C. $\begin{pmatrix} 10 & 14 \end{pmatrix}$

    D. $\begin{pmatrix} 3 \\ 7 \end{pmatrix}$

    E. $\begin{pmatrix} 10 \\ 7 \end{pmatrix}$

54. Find the determinant of the matrix. $\begin{pmatrix} 3 & 2 \\ 4 & 1 \end{pmatrix}$

    A. -7
    B. -5
    C. 2
    D. 5
    E. 9

55. What is the resulting matrix for $4\begin{pmatrix} 1 & 0 \\ 2 & 3 \end{pmatrix} + 2\begin{pmatrix} 2 & 3 \\ 1 & 0 \end{pmatrix}$?

A. $\begin{pmatrix} 6 & 2 \\ 4 & 2 \end{pmatrix}$

B. $\begin{pmatrix} 8 & 6 \\ 10 & 12 \end{pmatrix}$

C. $\begin{pmatrix} 4 & 0 \\ 8 & 12 \end{pmatrix}$

D. $\begin{pmatrix} 3 & 2 \\ 4 & 1 \end{pmatrix}$

E. $\begin{pmatrix} 4 & 6 \\ 2 & 0 \end{pmatrix}$

Complex Numbers

Complex numbers are in the form of z = a + bi. The distance, or modulus, is given by

$$|z| = |a + bi| = \sqrt{a^2 + b^2}.$$

The midpoint of the line segment joining the points (a + bi and s + ti) in the complex plane can be found by the following:

$$\text{Midpoint}: \frac{a+s}{2} + \frac{b+t}{2}i$$

56. What is the distance between the points (1 + 4i) and (3 - 2i) on the complex plane?

A. $\sqrt{13}$
B. 5
C. $\sqrt{17}$
D. $2\sqrt{10}$
E. 15

> The distance between the points is really the application of the Pythagorean Theorem. Draw out the points on the plane and you will see.

57. What is the midpoint between the points (1 + 4i) and (3 - 2i) on the complex plane?

    A. 2 - 3i
    B. 1 - i
    C. 2 + i
    D. 4 + 2i
    E. 3 + 4i

Vectors

A vector has both magnitude and direction.

58. What is the result of adding the vectors a = (3, 6) and b = (2, -5)?

    A. (1, 11)
    B. (5, 11)
    C. (5, 2)
    D. (5, 1)
    E. (6, 3)

Algebra

The Algebra subcategory accounts for between 12 and 15% of the total questions on the ACT.

We will work the following types of problems in this section:
- Evaluating Expressions
- Writing Expressions
- Combine Like Terms
- Distributive Property
- Simplify Variables
- FOIL
- Factoring
- Solving Equations
- System of Equations
- Slope
- Equation of a Line
- Graph of Line

- Inequalities
- Graph of Inequality
- Circles
- Distance

Evaluating Expressions

If asked to evaluate an expression, look for given values to plug in for variables and solve the equation.

59. If x = 3 and y = 4, what is the value of z in the following expression: z = 2x + 3y?
    A. 7
    B. 9
    C. 18
    D. 21
    A. 30

60. The ideal gas law is the equation of state of a hypothetical ideal gas. The ideal gas law is often written as PV=nRT. Solve this equation for T, the temperature of the gas.
    A. $T = PVnR$
    B. $T = \dfrac{PV}{nR}$
    C. $T = \dfrac{nR}{PV}$
    D. $T = \dfrac{PnR}{V}$
    E. $T = \dfrac{P}{VnR}$

> Make sure you do not get confused by lots of words and variables in the question, but focus on figuring out what specifically the question is asking you to solve. (We really don't care that this is the ideal gas law...just solve for T.)

61. Which equation represents a in terms of b for 2a - 8b = 4?
    A. 8b + 4
    B. 2b
    C. -4b - 2
    D. 4b + 2
    E. $\dfrac{b}{2}$

62. If (x,y) O (a,b) = xb - ya, what is (1,2) O (3,4)?

    A. -2
    B. 2
    C. 3
    D. 4
    E. 6

> This example uses a special operation symbol. This is used to match up the variable expression with the values as you plug into the expression. Many different operation characters can be used, so do not let that part confuse you.

63. The formula for exponential decay is given by the following: $y = A(1 - r)^t$, where y is the new value, A is the original value, r is rate as a decimal, and t is time in years. If the original value is 35, the rate is 12%, and the time is 5 years, what is y, to the nearest tenth?

    A. 13.2
    B. 15.9
    C. 18.5
    D. 22.7
    E. 30.1

> Do not allow the complexity of the equation to scare you. Determine the main task of the question, and if it is to plug in or solve for a variable, do that.

Writing Expressions

Being able to convert a word problem into an algebraic expression is an important skill you will need for this test. You should review key words, such as "sum", "difference", "product", and other terms, that will help you determine how the expressions should be written.

It is common to see a fixed and variable rate question on the test. There are many examples of real-world costs that may be used in this type of question.

64. Write an expression for 3 more than 2 times a number.

    A. 3 + 2x
    B. 2 + 3x
    C. 2 + 3
    D. (2 + 3)x
    E. (2 + x)(3x)

> When you check through your work at the end, try to solve the problem in a different way to confirm you answered correctly.

65. Emilee designs websites for customers. Her fees include a $50 consultation fee plus a $30/hour design fee. How many hours did she charge on a $200 bill?

    A. 2 ⅓
    B. 4
    C. 5
    D. 6 ⅔
    E. 8

> A *fixed cost* is one that does not change based on the number of occurrences of whatever you are calculating in the problem. The *variable cost* is dependent on the number of occurrences of the variable.

66. The sum of two numbers is 12. Their difference is 6. What is their product?

    A. 9
    B. 15
    C. 27
    D. 30
    E. 42

> Recognize the easy questions for you. Don't overthink these. Answer these and move on.

67. If x and y vary as given in the chart below, which of the following equations represents this relationship?

x	0	1	2	3	4
y	1	6	11	16	21

    A.   y = -5x + 1
    B.   y = x + 5
    C.   y = 5x + 1
    D.   y = 5x + 6
    E.   y = 21x

68. For 2 consecutive integers, the result of doubling the smaller integer and adding to the larger integer is 76. What are the 2 integers?
    A.   3 and 4
    B.   12 and 15
    C.   25 and 26
    D.   30 and 31
    E.   34 and 35

69. Regan is selling t-shirts for her summer business. She can rent a stand at the mall for $100/month. Each t-shirt costs her $2 to make, but she can sell one for $15. If she runs her business for 3 months at the mall, what is the expression that represents her profit when *x* t-shirts are produced and sold?
    A.   13x - 300
    B.   300 - 13x
    C.   15x - 100
    D.   13x
    E.   13x - 100

70. If $2y = x^2 + 4$, then what is $-y$ equivalent to?
    A. $\frac{(-x)^2}{2} - 2$
    B. $\frac{-x^2}{2} - 2$
    C. $\frac{x^2}{2} - 2$
    D. $\frac{-x^2}{2} + 4$
    E. $\frac{-x^2}{2} - 4$

Combining Like Terms

When you are working with variables (x, y, z, etc.), the first way to attempt to simplify terms is to determine if you are working with the same variables. For example, when you are adding x and y, that's as far as you can combine those terms, because x and y are different variables.

71. What is $(3x + 2y - z) - (4x - 5y + 3z)$ equivalent to?

    A. $-x - 3y + 2z$
    B. $-x + 7y - 4z$
    C. $12x^2 + 10y^2 - 3z^2$
    D. $7x + 7y + 4z$
    A. $7x - 3y - 4z$

72. What must be added to $x^2 + 5x$ so that the sum is $3x^2 - 2x + 4$?

    A. $3x - 2$
    B. $x^2 - 2x + 4$
    C. $2x + 4$
    D. $x^2 - 3x + 4$
    E. $2x^2 - 7x + 4$

> Combining like terms means looking for variables that are the same, and raised to the same power, and simplifying them.

Distributive Property

When we use the distributive property, we are distributing the single term, outside the expression, through to each term within the expression. In other words, we multiply the outside term by each of the terms that are being added together within the parentheses. In this section, we will look at one question, but this skill can be used throughout different questions on the test.

73. Evaluate the expression $-3xy(x^2 - 2xy^3)$.

    A. $-3x^3y - 6x^2y^4$
    B. $-3x^2y - 6x^2y^3$
    C. $-3x^3y + 6x^2y^4$
    D. $-3x^2y - 6x^2y^3$
    E. $-3x^3y$

Simplify Variables

When we simplify variables, we are finding an equivalent expression, but in a simpler form. To simplify variables, you may have to combine like terms, factor, simplify exponents, or perform a number of techniques that you have learned.

74. Simplify the expression $(x - 1)/(x^2 - 2x + 1)$.

   A. $\dfrac{1}{x}$

   B. $\dfrac{1}{x - 1}$

   C. $\dfrac{1}{x + 1}$

   D. $\dfrac{x - 1}{x + 1}$

   E. 13

75. For all positive real numbers x, which expression is equivalent to $\dfrac{\dfrac{x^{18}}{x^7}}{\dfrac{1}{x^3}}$?

   A. 1
   B. x^3
   C. x^8
   D. x^{14}
   E. x^{33}

> The goal is to get the expression in the form that can no longer be simplified.

FOIL

The FOIL method is used to multiply together two binomials. The acronym FOIL stands for First, Outer, Inner, Last. This is a technique to help you to remember to multiply each one of the components by each other in solving this type of problem.

76. The expression $(x^2 + 6)(x - 3y)$ is equivalent to?

   A. $x^2 - 3xy + 6x - 18y$
   B. $x^3 - 18y$
   C. $x^2 - 3x^2y + 6x - 18y$
   D. $x^3 - 3x^2y + 6x - 18y$
   E. $x^3 - 3x^2y - 6x - 18y$

77. What is $(z + 3x)^2$ equivalent to?

    A. $z^2 + 3$
    B. $z^2 + 9x^2$
    C. $z + 3x$
    D. $z^2 + 6xz + 9x^2$
    E. z

> Another way that I was taught to remember to multiply each term in these types of problems is by saying "first times first, first times second, second times first, and second times second".

Factoring

When we are factoring in Algebra, we want to find what we can multiply together to get an expression. We start with an expression and we want to break it out into the "mini" expressions that multiply together to get the original expression.

78. What is a factored form of the expression $x^2 + x - 12$?

    A. $(x - 4)(x - 3)$
    B. $(x - 4)(x + 3)$
    C. $(x + 4)(x - 3)$
    D. $(x + 4)(x + 3)$
    E. $(x + 6)(x - 2)$

79. What values are the solution for $x^2 + x = 6$?

    A.   -3 and 2
    B.   6 and 0
    C.   3 and -2
    D.   6 and 1
    E.   1 and 0

> If you find that you are spending more than 30 seconds on a problem (and not near finishing), mark that #, move on to another problem, and come back to it at the end.

Solving Equations

To solve algebraic equations, use the skills such as combining like terms, simplifying, and moving the variable you are solving for on one side.

80. What is x equal to in the equation $3^{x+1} = 9^{2x+2}$?

    A. -2
    B. -1
    C. 1
    D. 2
    E. 6

> If you have the same bases, the exponents are equivalent in equations.

81. If $\dfrac{2}{\sqrt{5}} = \dfrac{a\sqrt{5}}{3}$, what is a?

    A. $\sqrt{3}$
    B. 2
    C. $\dfrac{6}{5}$
    D. 4
    E. 5

82. What are the real solutions to the following equation: $|x| - 3|x| + 10 = 0$?

    A. 0, 5
    B. -2, 5
    C. -1, 4
    D. -5, 5
    E. 3, 10

> It is important to remember that for absolute value, the terms within the absolute value can be positive or negative.

System of Equations

A system of equations is two or more equations with the same unknowns or variables. As long as you have at least the same number of unique equations as you do unknowns, you can solve for the unknown variables. There are several methods to use to solve a system of equations, including using matrices, solving for one variable in terms of the other and plugging in, and adding or subtracting equations to eliminate variables. The method you choose to use is based on your understanding, timing in working the problems, and ease of application in the system given.

83. What is the value of x and y if given the following equations:
$$x + 3y = 6$$
$$2x + y = 3?$$

   A. $x = 3$ and $y = \frac{9}{5}$

   B. $x = \frac{3}{5}$ and $y = \frac{9}{5}$

   C. $x = 3$ and $y = 9$

   D. $x = \frac{2}{5}$ and $y = \frac{9}{5}$

   E. $x = 5$ and $y = \frac{9}{5}$

84. For what value of *a* would the following system of equations have an infinite number of solutions?
$$3x - y = 12$$
$$12x - 4y = 8a$$

   A. 3
   B. 4
   C. 6
   D. 12
   E. 48

Slope

The slope of a line in the coordinate plane is defined as the change in y over the change in x. Slope can also be thought of as the "rise over the run". Slope is usually represented by "m" as in the slope-intercept equation,
$y = mx + b$.

85. What is the slope of the line through (3,2) and (-4, 5) in the standard (x,y) coordinate plane?

   A. $-\frac{1}{2}$

   B. $-\frac{3}{7}$

   C. 1

   D. $\frac{3}{7}$

   E. $\frac{1}{2}$

86. Given the equation 4x - 3y = 7, what is the slope of the line perpendicular to this line?

    A. $-\dfrac{7}{3}$

    B. $-\dfrac{3}{4}$

    C. $\dfrac{4}{3}$

    D. 3

    E. 4

87. What is the slope of the line parallel to 8x - 3y = 1?

    A. -8

    B. $-\dfrac{3}{8}$

    C. $\dfrac{8}{3}$

    D. 3

    E. 8

> Parallel lines have the same slope, and perpendicular lines have slopes that are negative reciprocals.

88. What is the slope of the line perpendicular to 2y - 4x = 6?

    A. -4

    B. -2

    C. -½

    D. 2

    E. 4

> A positive slope is a line increasing from left to right (↗). A negative slope indicates a decrease from left to right (↘).

Equation of Line

For an equation of a line, we often think of the slope-intercept form, y = mx + b. In this equation, we have the variables x and y. The m represents the slope of the line. The b is the y-intercept, the point on the graph where x = 0 and the line crosses the y-axis.

89. The graph of y = 2x + 4 passes through (8a,12) in the standard (x,y) coordinate plane. What is the value of a?

    A. -2
    B. $\frac{1}{2}$
    C. $\frac{3}{2}$
    D. 2
    E. 4

90. Line t in the standard (x,y) coordinate plane has equation x = 4 and intersects line u given by equation y = x + 1. What is the point of intersection of lines t and u?

    A. 0
    B. 1
    C. 4
    D. 5
    E. 7

If you are solving a problem and get stuck, try to plug in some of the solutions to see what works.

Graph of Line

Using what we know about points, slope, and an equation of a line, we can graph a line on the coordinate axis.

91. The point (1,3) is shown in the standard (x,y) coordinate plane. If there is a line that passes through (1,3) with slope ⅓, what is another point on that line? Use the plane below.

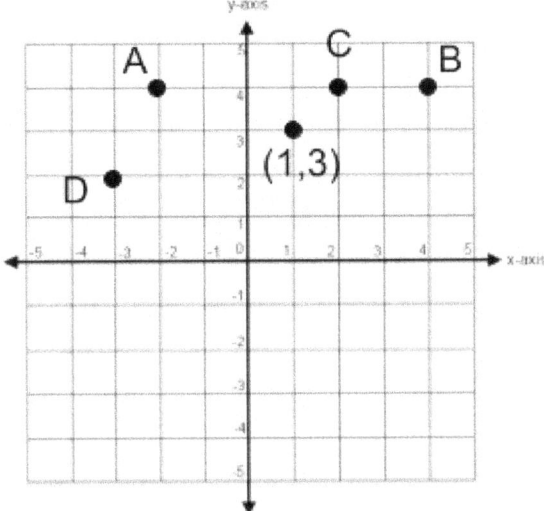

A. A
B. B
C. C
D. D

92. Which of the following is the graph of the equation x + 2y = 8 in the standard (x,y) coordinate plane?

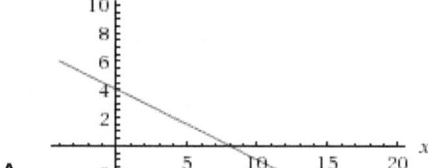
A.

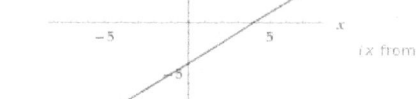
B.

C.

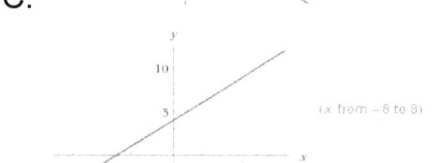
D.

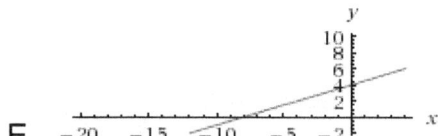
E.

Inequalities

When a math equation has two expressions that are not equal, they are inequalities. One expression is either greater than (>) or less than (<) the other expression. Do not forget about the ≥ and ≤ cases as well.

93. At a manufacturing facility, the tolerance for widget length, l, is represented by the inequality |l - 1.5| ≤ 0.01. What is the range of lengths the widget can be to pass inspection?

    A.   l ≥ 0.01
    B.   1.49 ≤ l ≤ 1.51
    C.   1.51 ≤ l
    D.   1.40 ≤ l ≤ 1.60
    E.   1.49 ≥ l ≥ 1.51

94. Simplify the inequality 5(x + 1) > 2(x - 1). (On your own...Graph on a number line.)

   A. $x > -\frac{7}{3}$

   B. $x < -\frac{7}{3}$

   C. $x > \frac{7}{3}$

   D. $x > \frac{3}{7}$

   E. $x < \frac{3}{7}$

Graph of Inequality

The graph of an inequality on the coordinate plane allows us to see the shaded representation on the region of the inequality, along with the boundary. A ≥ or ≤ inequality results in a solid line being the boundary, whereas > or < results in a dashed or dotted line as the boundary.

95. Is the shaded region of the graph y ≤ x + 1 or y ≥ x + 1?

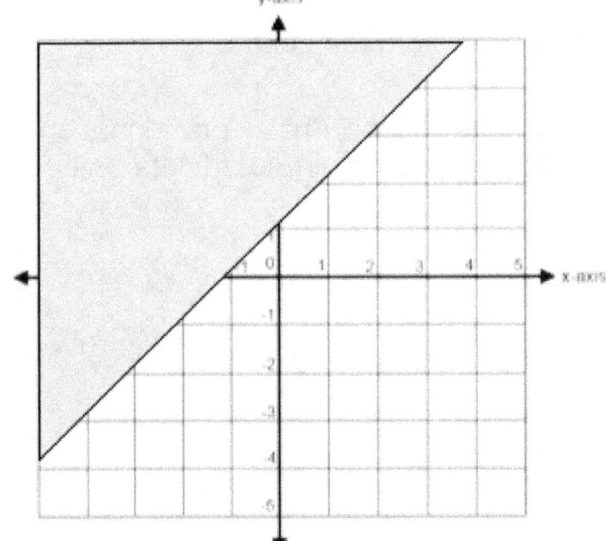

   A. y ≤ x + 1
   B. y ≥ x + 1
   C. Both
   D. Neither

> To determine the shaded region, you must look at the inequality and try a few points.

96. The graphs of $y = x^2$ and $y = x$ are shown below. What real values of x satisfy the inequality $x^2 < x$?

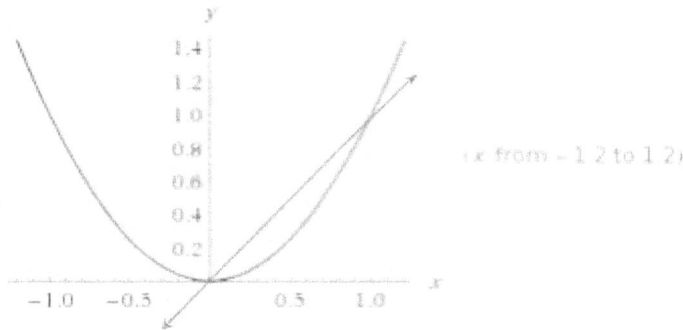

   A. x > 1
   B. x < 0
   C. x > .5
   D. 0 < x < 1
   E. x < .5

Circles

A circle is a curve that always has the same distance, r, from the center.

97. A circle in the standard (x,y) coordinate plane has an equation of $(x + 3)^2 + (y - 2)^2 = 9$. What are the radius of the circle, in coordinate units, and the coordinates of the center of the circle?

   A. 9, (3, -2)
   B. 3, (3, -2)
   C. 3, (-3, 2)
   D. 9, (-3, 2)
   E. 81, (-3, 2)

98. What is the equation of the circle in the graph?

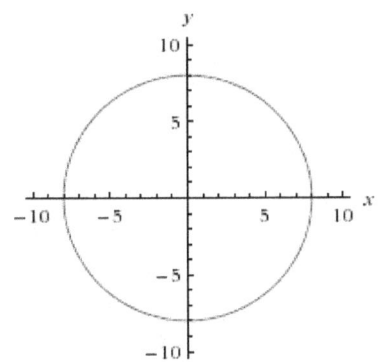

A. $(x - 2)^2 + y^2 = 8$
B. $x^2 + y^2 = 64$
C. $x^2 + y^2 = 8$
D. $(x - 5)^2 + (y - 5)^2 = 25$
E. $x^2 + y^2 = 25$

> The standard equation of a circle is
> $(x - h)^2 + (y - k)^2 = r^2$.
> In this equation, (h, k) is the center and r is the radius.

99. What is the center of the circle given by the equation $(x - 3)^2 + (y + 2)^2 = 25$?

A. (-1, 2)
B. (-3, 2)
C. (3, -2)
D. (4, -1)
E. (5, -2)

Distance

In the coordinate plane, we may want to look at the distance, or how far apart, between points. The equation for this often gives students a hard time, but remember that you are looking for how far the x-values are from one another and how far the y-values are from one another. We square the result because we do not want a negative difference. Then, add up the differences, and take the square root.

$$\text{Distance} = \sqrt{(x_2 - x_1)^2 + (y_2 - y_1)^2}$$

100. What is the length of the line segment with endpoints (3,7) and (-2, 4) in the standard (x,y) coordinate plane?

    A. 5
    B. $\sqrt{34}$
    C. 7
    D. $\sqrt{53}$
    E. 8

Functions

The Functions subcategory accounts for between 12 and 15% of the total questions on the ACT.

A function gives a relationship where for every input there is exactly one output. A function is often written as f(x), but it can be written with other terms than "f", so do not let that confuse you. The term "f(x)" means the function, f, written in terms of x. To keep things simple, you can also think of f(x) as y. If you are given f(x) = x + 3, you can also think y = x + 3.

We will work the following types of problems in this section:
- Arithmetic and Geometric Sequences
- Logs
- Reading Graphs
- Solving Functions
- Asymptotes
- Even/Odd Functions
- Transformations

Arithmetic/Geometric Sequence

Working questions related to arithmetic and geometric sequences requires finding a pattern. For arithmetic sequences, each term is equal to the previous term plus a common difference, a constant. In a geometric sequence, each term in the sequence is multiplied by the same number. The problems on the test can be as straight-forward as finding the terms in a sequence, to more challenging problems of finding terms many places farther along in the sequence and summing specific terms in the sequence. Knowing the relationships in the sequences will help you solve all these types of questions.

101. What is the next term in the arithmetic sequence {1, -3, -7, -11, ?}?

    A.   -20
    B.   -17
    C.   -15
    D.   5
    E.   7

102. What is the missing term in the following geometric sequence: $\{1, \frac{1}{2}, \frac{1}{4}, ?, \frac{1}{16}, \frac{1}{32}\}$?

    A. $\frac{1}{14}$
    B. $\frac{1}{12}$
    C. $\frac{1}{10}$
    D. $\frac{1}{8}$
    E. $\frac{1}{6}$

> If you find yourself missing alot of the same type of problems, focusing on those and really learning how to solve them could result in making a big impact by getting more questions correct on the next try.

103. What is the sum of the first 5 terms of the arithmetic sequence in which the 7th term is 6.5 and the 12th term is 10.25?

    A.   3.25
    B.   5
    C.   16.75
    D.   17.5
    E.   21.25

Logs

Logs are the opposite, or inverse, of exponents. Logs are the power that a number must be raised to get some number. You can think of logs as undoing exponents.

104. What is the value of $\log_3 27$?

    A. 2
    B. 3
    C. 6
    D. 9
    E. 13

> The relationship between logs and exponents is as follows:
> $\log_b y = x$
> $b^x = y$

105. What is the real value of x in the equation $\log_2 16 - \log_2 4 = \log_3 x$?

    A. 6
    B. 9
    C. 12
    D. 15
    E. 20

> There are rules related to logs that you should be familiar with and may encounter on the test.
> - $\log_b (mn) = \log_b m + \log_b n$
> - $\log_b (m/n) = \log_b m - \log_b n$
> - $\log_b (m^n) = n \log_b m$

106. For what real value of x is $\log_{x+1}(x^2 + 5) = 2$?

    A. -2
    B. -1
    C. 0
    D. 1
    E. 2

107. When is $f(x) = \log_b(x + 3)$ undefined?

    A. $x \leq -3$
    B. $x \geq -3$
    C. $x \leq 0$
    D. $x < -3$
    E. $x > -3$

> Hint: $\log_b c$ is undefined when $c \leq 0$.

Reading Graphs

It is important to be able to read and use a graph that is given to you. Be familiar with the axes, units, and any other details given that will help you to answer the questions.

108. Widgets x and y are manufactured daily on a machine. The daily machine capacity is 10 (x + y ≤ 10). If the business earns $200 for every x produced and $300 for every y produced, what is the maximum profit they can earn from x and y widgets in a day?

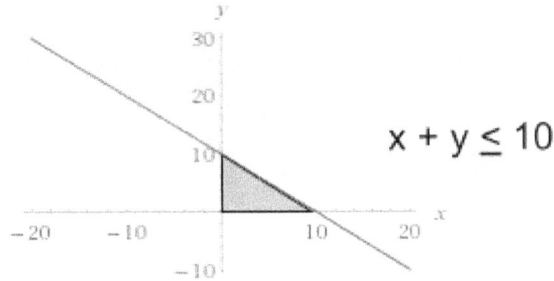

x + y ≤ 10

- A. $300
- B. $500
- C. $2000
- D. $3000
- E. $5000

Solving Functions

In this section we have examples for plugging in values for x in the function. You can think "find x in the expression and plug the value in for x". There are several more challenging problems to try, but these types are often seen on the test.

109. If $f(x) = \sqrt{x + 1}$ and $g(x) = 2x - b$ and in the standard (x,y) coordinate plane y = f(g(x)) passes through (5,4), what is the value of b?

- A. -8
- B. -5
- C. 0
- D. 4
- E. 5

110. Find f(-2) if $f(x) = -x^2 + 4x$.

    A.   -12
    B.   -4
    C.   4
    D.   6
    E.   12

> Eliminate answer choices as you can. Then, choose from the answers remaining.

111. Find $f(g(x))$ if $f(x) = x^2 + 3x$ and $g(x) = x + 7$.

    A. $x^2 + 17x + 70$
    B. $x^2 + 3x + 49$
    C. $4x + 7$
    D. $x^2 + 14x + 49$
    E. $x + 7$

112. A function H is defined as follows:

$$H(x) = x^2 + 3x - 1, \text{ for } x > 1$$
$$H(x) = -x^2 - 2x + 1, \text{ for } x \leq 1$$

What is the value of H(2)?

    A.   -2
    B.   -1
    C.   1
    D.   9
    E.   10

> A function is often written as f(x), but it can be written with other terms than "f", so do not let that confuse you.

Asymptotes

An asymptote is a line that a curve approaches as it heads towards infinity. There are 3 types of asymptotes: vertical, horizontal, and oblique (or slant).

To find a vertical asymptote, set the denominator equal to 0.

The function has a horizontal asymptote of y=0 if the power of x is larger in the denominator. If the powers are the same in the numerator and denominator, the horizontal asymptote is equal to the coefficients of the highest powered variables.

For an oblique, or slant, asymptote, the power of x is larger in the numerator and you must use long division. The "whole number" portion of the division result is the oblique asymptote.

113. What type of asymptote does the function $f(x) = (x^2 + 3x + 2)/(x-2)$ have?

    A.   Vertical Only
    B.   Horizontal Only
    C.   Oblique Only
    D.   Vertical and Horizontal
    E.   Vertical and Oblique

114. What is/are the asymptote(s) of the function $f(x) = 1/(x-2)$?

    A.   x = 0 only
    B.   x = 2 only
    C.   x = 2 and y = 0
    D.   x = 0 and y = 0
    E.   none

Even/Odd Functions

It is also useful to be able to determine if a function is even, odd, or neither. A function is even if $f(-x) = f(x)$. Graphically, we can determine this if the graph is symmetric to the y-axis. This indicates that the function is even.

A function is odd if $f(-x) = -f(x)$. Again, we can view a graph of the function, and if the graph is symmetric about the origin, it is odd. The 180° test is helpful for this determination. Imagine rotating the graph by 180°. If the graph looks the same, it's symmetric about the origin and an odd function.

If both of these algebraic or graph checks fail, then the function is neither even nor odd.

115. Below is the graph of y=cosx. Is this function even, odd, or neither?

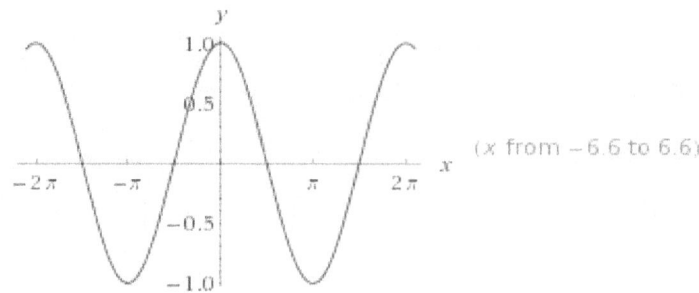

- A. Even
- B. Odd
- C. Neither
- D. Even and Odd
- E. Unable to determine

116. A function f is an even function if and only if f(-x) = f(x) for all x in the domain of f. Which function graphed below is NOT an even function?

A.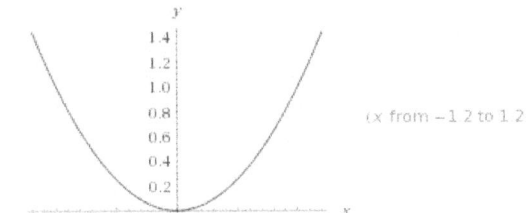
(x from −1.2 to 1.2)

B.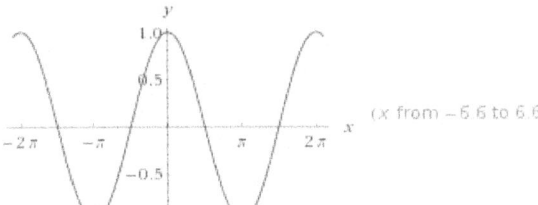
(x from −6.6 to 6.6)

C.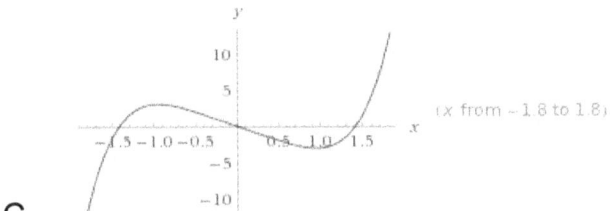
(x from −1.8 to 1.8)

D.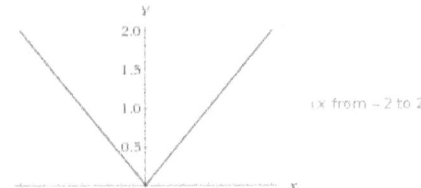
(x from −2 to 2)

E.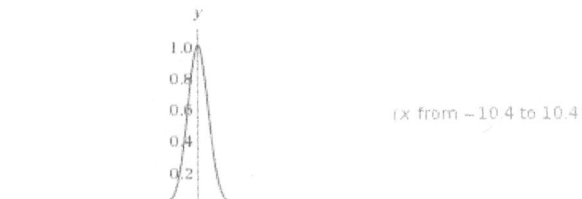
(x from −10.4 to 10.4)

117. Below is the graph of y=sinx. Is this function even, odd, or neither?

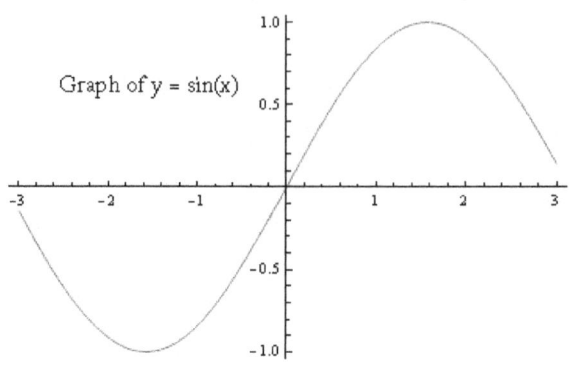

- A. Even
- B. Odd
- C. Even and odd
- D. Neither
- E. Unable to determine

Transformations

The general form of the trigonometric functions can take on many different shapes and forms and this is referred to as transformations. Refer to the Formulas section for the modifications to the general form that you should be aware of in transformation questions.

118. Given $y_1(t) = a_1\sin(b_1 t)$ and $y_2(t) = a_2\cos(b_2 t)$, how do b_1 and b_2 compare?

$y_1(t) = a_1\sin(b_1 t)$

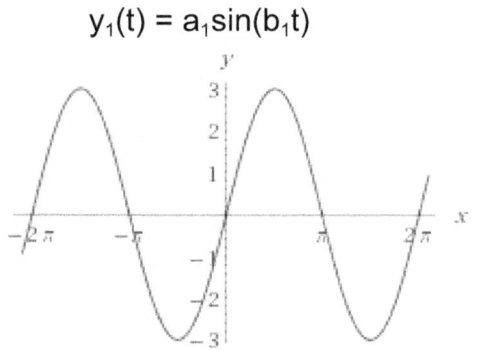

$y_2(t) = a_2\cos(b_2 t)$

- A. $b_1 > b_2$
- B. $b_2 > b_1$
- C. $b_2 = b_1$
- D. Unable to determine

119. What is an equation for a cos graph with a period of 2π, an amplitude of 2, and an upward shift of 3 units?

    A.   y = 2 cos (2πx) - 3
    B.   y = -2 cos (2πx) - 3
    C.   y = 3 cos (x) + 3
    D.   y = -2 cos (x) + 3
    E.   y = -2 cos (2x) + 3

Geometry

The Geometry subcategory accounts for between 12 and 15% of the total questions on the ACT. Expect to see the (x, y) coordinate plane. Graphing and interpreting equations are tested in these questions.

We will work the following types of problems in this section:
- Quadrants
- Planes
- Midpoint
- Parallel Lines
- N-Sided Polygons
- Surface Area
- Angles
- Pythagorean Theorem
- Similar Triangles
- Volume
- Circles
- Circle - Central Angles and Arcs
- Inscribed Angles
- Conic Sections
- Sine
- Cosine
- Tangent
- Trigonometric Identities
- Law of Sines
- Law of Cosines

Quadrants

The (x, y) coordinate plane is divided into 4 quadrants. They are usually labeled I, II, III, and IV. Quadrant I is in the upper right, with quadrants II, III, and IV following in order in a counter-clockwise direction.

120. If point C has a non-zero x-coordinate and a non-zero y-coordinate and the coordinates have the same signs, then point C must be located in which of the 4 quadrants labeled below?

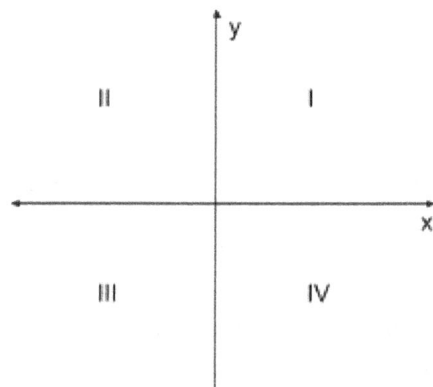

- A. I only
- B. II and III
- C. I and III
- D. I and IV
- E. III only

121. What are the quadrants of the standard (x,y) coordinate plane below that contain points on the graph of the equation x - y = 2?

- A. III only
- B. I and III only
- C. I, II, III, and IV
- D. II and IV only
- E. I, III, and IV only

Planes

A plane is a flat, 2-dimensional surface that extends infinitely far...forever. When we draw a plane, we have to give it an edge, because we cannot draw something infinite in length. But, you have to imagine it extending out as far as you can imagine and beyond! It has no thickness. Parallel planes never intersect.

122. Two distinct planes that are not parallel intersect to form which of the following?

- A. point
- B. angle
- C. line
- D. line segment
- E. ray

> If you miss a question, rework it until you get it correct and fully understand it. You probably will not see the exact problem again, but if you understand the problem, you can apply that knowledge to a similar problem.

Midpoint

This section goes a little more in-depth with midpoint. Previously, we calculated the midpoint of a line. In this section, we practice the same skill, but we may be solving for different unknowns. Use what you know about midpoint to solve these types of questions.

123. In the standard (x,y) coordinate plane, the midpoint of AB is (4,1) and A is located at (2,3). If (x,y) are the coordinates of B, what is the value of x - y?

    A. -1
    B. 0
    C. 4
    D. 6
    E. 7

> Hint: Work "backwards" from the midpoint to find B. Then use those coordinates to answer the question.

124. In the standard (x,y) coordinate plane, point M with coordinates (6,1) is the midpoint of AB and B has coordinates (7,3). What are the coordinates of A?

    A. (5, -1)
    B. (1, 3)
    C. (1, -1)
    D. (6, -1)
    E. (5, 1)

Parallel Lines

Lines are parallel if they never touch. They never intersect. Oftentimes when discussing parallel lines, transversals will come up. A transversal is a line that cuts across two or more parallel lines. From this cut, angles are formed, and there are relationships between these angles.

125. In the figure below, lines l and m are parallel. Line t is a transversal that intersects both l and m. Give the sets of angles with equivalent angle measures.

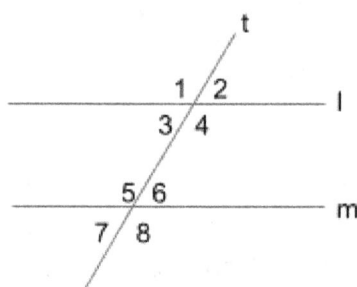

A. {1, 2}, {3, 4}
B. {5, 7}, {1, 3}
C. {1, 3, 5, 7}, {2, 4, 6, 8}
D. {1, 4, 5, 8}, {2, 3, 6, 7}
E. {1, 2, 3, 4}, {5, 6, 7, 8}

126. In the figure below, lines l and m are parallel and intersected by lines s and t. Given the angle measures below, what is the measure of angle x?

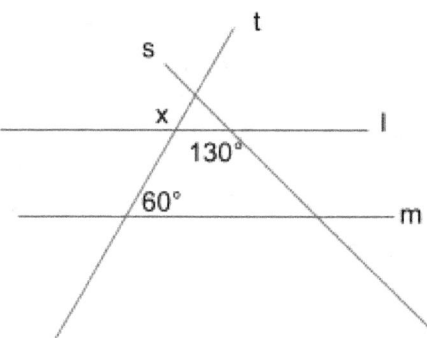

A. 30°
B. 60°
C. 90°
D. 120°
E. 130°

N-Sided Polygons

Polygons are made of straight lines and the shapes are closed. In a regular polygon, all the sides are the same length and the angles are the same measure.

127. If 2 angles of a pentagon measure 70° and 100°, what is the sum of the other angles of the pentagon?

    A. 70°
    B. 100°
    C. 170°
    D. 370°
    E. 540°

> Be familiar with the prefixes related to the number of sides of a polygon. For example, for "tri", think "3".

Surface Area

We will take a look at the surface area for figures, including prisms and pyramids, in this section. For figures with curves, the calculation is a little more complex. The examples we have seen on the actual tests include the surface area of a right circular cylinder and that formula has been given on the test. It is still a good idea to be familiar with this formula in case it is removed from the test question in the future.

128. If the surface area of a square cube is 96 in², how long, in inches, are the edges of the cube?

    A. 4
    B. 9.8
    C. 16
    D. 24
    E. 48

> Surface Area of a Prism = 2A + pl, where A is the area of the bases, p is the perimeter of the base, and l is the height of the prism.

129. What is the total surface area of this right circular cylinder, in square inches? (Note: The total surface area of a cylinder is given by $2\pi r^2 + 2\pi rh$ where r is the radius and h is the height.)

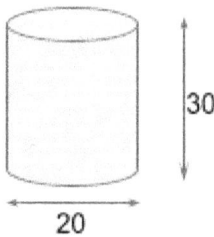

A. 650π
B. 800π
C. $800\pi^2$
D. 850π
E. $850\pi^2$

> Notice the SA formula for the cylinder is still SA = 2A + pl. $2\pi r^2$ is 2 times the area of the base. $2\pi r$ is the circumference of the circular base, which is the perimeter. The height is h.

Angles

An angle is formed by two lines or rays. At the point of intersection, an angle is formed. Figures contain angles. You will encounter questions on the test related to the relationship of the angles in figures.

130. In the isosceles trapezoid ABCD, AB is parallel to DC, ∠BDC measures 35°, and ∠BCA measures 25°. What is the measure of ∠DBC?

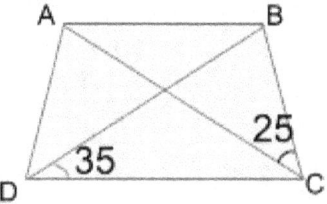

A. 25°
B. 35°
C. 85°
D. 120°
E. Cannot be determined

131. Given the angle measures below, if points O, P, R, and S are collinear, what is the measure of ∠QRS?

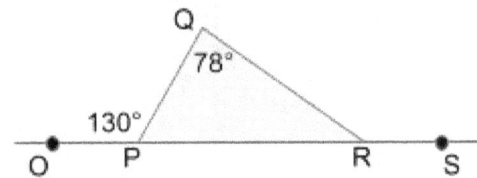

   A. 78°
   B. 102°
   C. 128°
   D. 130°
   E. Cannot be determined

> Create a formula sheet and review it often. (I have included one in the back, but it is a great idea to make your own too!)

132. For all △'s ABC where ∠A < ∠B, how do the lengths of BC and AC compare?

   A. BC ≥ AC
   B. BC = AC
   C. BC > AC
   D. BC < AC
   E. Cannot be determined

Pythagorean Theorem

The Pythagorean Theorem is a fundamental relationship between the three sides of a right triangle. A right triangle is composed of a hypotenuse, which is the side opposite the 90° angle, and two other sides. The theorem states that the hypotenuse squared is equal to the sum of the squares of the two other sides. ($c^2 = a^2 + b^2$).

133. In the right triangle pictured below, what is the length of the hypotenuse?

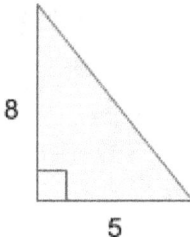

- A. $\sqrt{13}$
- B. $\sqrt{89}$
- C. 13
- D. 54
- E. 89

134. Given an isosceles right triangle with one leg = 5 units, what is the length of the hypotenuse?

- A. 10
- B. 5
- C. $5\sqrt{2}$
- D. 10
- E. Unable to determine

> Remember that the hypotenuse is longer than each of the other 2 sides. Knowing this may help you eliminate some answer choices.

135. What is the length of a leg of a right triangle with hypotenuse of 13 and one leg 11?

- A. 1
- B. $2\sqrt{3}$
- C. 3.2
- D. $4\sqrt{3}$
- E. 5

136. What is an expression for *a* in terms of *b* and *c*?

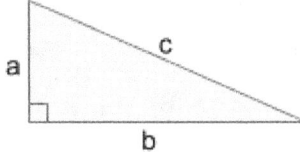

- A. $\sqrt{b^2 + c^2}$
- B. $\sqrt{(-b^2 + a^2)}$
- C. $\sqrt{(-a^2 + c^2)}$
- D. $\sqrt{(-c^2 + b^2)}$
- E. $\sqrt{(-b^2 + c^2)}$

Similar Triangles

Triangles are similar if the corresponding sides are in proportion and the corresponding angles are congruent. One triangle may be smaller or larger than the other, but the sides are in proportion, almost as if you have shrunk or enlarged one from the other. The triangles may be in the same orientation or one may be rotated somewhat. This does not matter as long as the side proportions and angle congruency hold true.

137. What is the perimeter of △PQR? △ABC ~ △PQR (Note: The symbol ~ means "is similar to".)

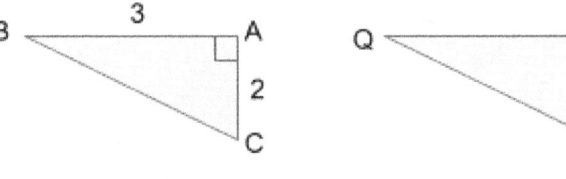

- A. $5 + \sqrt{13}$
- B. 9
- C. $15 + \sqrt{13}$
- D. $15 + 3\sqrt{13}$
- E. 32

138. In right triangle PQR, ST is parallel to PQ, and ST is perpendicular to QR at T, PR is 30 inches, ST is 3 inches, and TR is 4 inches. What is the length, in inches, of PQ?

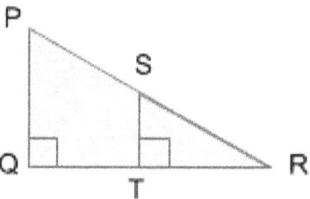

    A.   6
    B.   12
    C.   18
    D.   25
    E.   30

> Most numerical answer choices are in numerical order. When you need to plug in answer choices to try to solve a question, try the middle answer choice (C) first. If the middle answer choice is not the solution, based on your answer with "C", determine if you should try a higher value (D or E) or lower value (A or B) next. This technique may save you time by not having to go through all the answer choices.

Volume

Volume refers to the amount of space within a 3-dimensional figure. In a figure, when determining the volume, you must consider the length, width, and height of the figure. Depending on the shape of the figure, these components of volume may be measured in different ways.

139. Julie is packing items that she has sold online. She needs to find a box that her product will fit inside. One box she is looking at holds a volume of 4032 cubic inches. The box description says it has a length of 18" and a width of 16". How tall, in inches, is this box?

    A.   12"
    B.   14"
    C.   16"
    D.   18"
    E.   20"

140. The volume of a right cylinder is 37.68 cubic feet. If the height of the cylinder is 3 feet, what is the diameter, in feet? Use 3.14 for π.

   A. 2
   B. 3.7
   C. 4
   D. 6
   E. 15.2

> The units for volume are cubed because you are multiplying 3 measures of dimension.

Circles

Apply the definition of a circle to solve this question in the coordinate plane.

141. As shown in the standard (x,y) coordinate plane, P(3,7) lies on the circle with center (3,3) and radius 4 coordinate units. What are the coordinates of P after the circle is rotated 180° about the center of the circle?

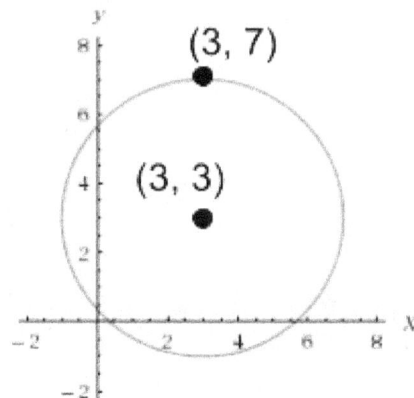

   A. (3, -.5)
   B. (3, -1)
   C. (7, 3)
   D. (-1, 3)
   E. (5.5, 0)

Circle - Central Angles and Arcs

A central angle is formed by the meeting of two radii of a circle. The center of the circle is a central angle's vertex. The portion of the circle's circumference that is formed by the intersection of the circle and the two radii (legs of the central angle) is an arc. The measure of the central angle and the degree measure of the arc are equivalent.

The length of the arc is a proportion of the circumference based on the measure of the central angle that forms the arc upon intercepting the circle.

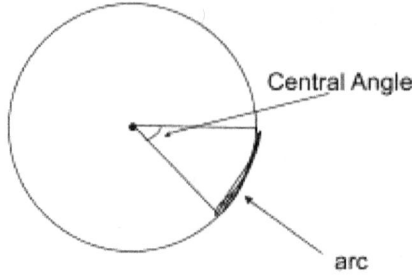

142. What is the measure of the central angle x?

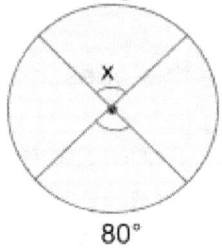

    A.   20°
    B.   40°
    C.   80°
    D.   160°
    E.   Unable to determine

143. Given a standard wall clock, as in the figure below, what is the arc measure, in degrees, of the arc between 3 and 5?

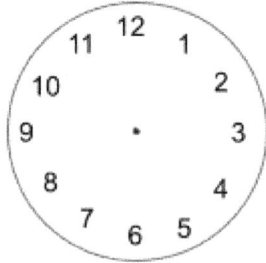

    A.   15
    B.   20
    C.   30
    D.   45
    E.   60

144. Given the figure below, what is the measure of the cut out arc length, in inches, to the nearest tenth?

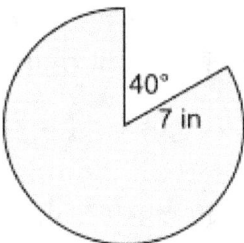

- A. 4.9
- B. 5.7
- C. 21.3
- D. 43.9
- E. 44.2

Inscribed Angles

An inscribed angle is an angle formed by 2 chords in a circle which have a common endpoint (vertex). The other two endpoints form an intercepted arc. The measure of the inscribed angle is half the measure of the corresponding central angle and arc measures.

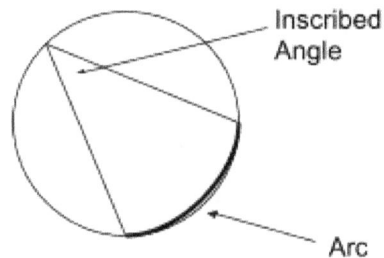

145. In the figure shown below, what is the degree measure of arc QR?

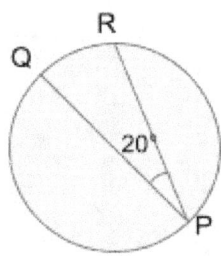

- A. 10°
- B. 20°
- C. 35°
- D. 40°
- E. 42°

Conic Sections

A conic section is the intersection of a plane and a cone. The resulting section may be a circle, ellipse, parabola, or hyperbola. When the plane touches the vertex, a point, line, or 2 intersecting lines can form. The form of the equation is as follows:

$$Ax^2 + Bxy + Cy^2 + Dx + Ey + F = 0$$

The following guidelines can be used to classify equations of conic sections:

$$\text{parabola: } Ax^2 + Dx + Ey = 0$$
$$\text{circle: } x^2 + y^2 + Dx + Ey + F = 0$$
$$\text{ellipse: } Ax^2 + Cy^2 + Dx + Ey + F = 0$$
$$\text{hyperbola: } Ax^2 - Cy^2 + Dx + Ey + F = 0$$

146. What conic section is the following equation classified as?

$$x^2 + 2x = -3y^2 - 4y + 36$$

- A. parabola
- B. circle
- C. ellipse
- D. hyperbola
- E. none of the above

Sine

The sine of an angle is another trigonometric function of a right angle that is equal to the ratio of the side opposite the angle and the hypotenuse.

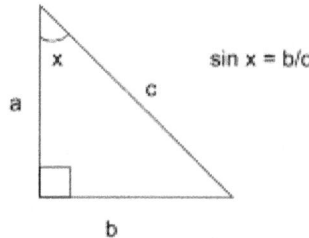

147. Find sin θ.

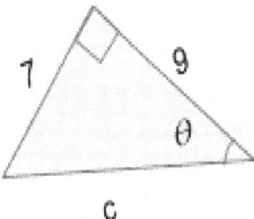

A. $\dfrac{c}{9}$

B. $\dfrac{9}{7}$

C. $\dfrac{7}{c}$

D. $\dfrac{7\sqrt{130}}{130}$

E. $\dfrac{\sqrt{130}}{7}$

As long as you are given 2 sides of a right triangle, you can use the Pythagorean Theorem to find the 3rd side, if needed.

148. Given AB and AC, what trigonometric expression gives the measure of ∠ABC?

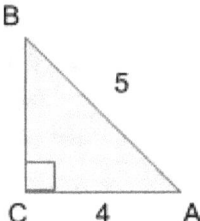

A. $\sin^{-1}\left(\dfrac{5}{4}\right)$

B. $\sin^{-1}\left(\dfrac{4}{5}\right)$

C. $\sin^{-1}\left(\dfrac{3}{5}\right)$

D. $\cos^{-1}\left(\dfrac{5}{4}\right)$

E. $\cos^{-1}\left(\dfrac{4}{5}\right)$

149. If sin θ = 0.8, what is the length of BD (x), to the nearest tenth?

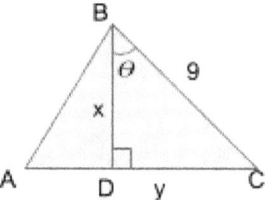

- A. 1.3
- B. 1.5
- C. 3.2
- D. 5.4
- E. 7.2

> Be prepared to work multi-step problems...use data from one result to get the final answer.

Cosine

In a right triangle, cosine is a trigonometric function giving the relationship of the adjacent side length of an angle to the hypotenuse. The cosine of an angle is equal to the adjacent side divided by the hypotenuse.

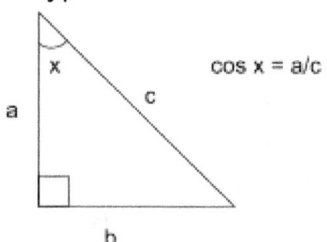

150. In the figure to the right, find cos θ.

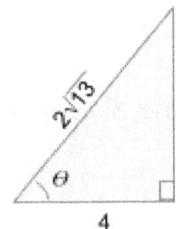

A. $\dfrac{2\sqrt{13}}{4}$

B. $\dfrac{2\sqrt{13}}{13}$

C. 4

D. $8\sqrt{13}$

E. $\dfrac{13\sqrt{2}}{4}$

> As long as you are given 2 sides of a right triangle, you can use the Pythagorean Theorem to find the 3rd side, if needed.

151. What expression gives the measure of ∠B?

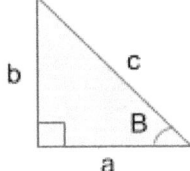

A. $\cos^{-1}\left(\dfrac{b}{c}\right)$

B. $\cos^{-1}\left(\dfrac{c}{a}\right)$

C. $\cos^{-1}\left(\dfrac{a}{c}\right)$

D. $\cos^{-1}\left(\dfrac{b}{a}\right)$

E. $\cos^{-1}\left(\dfrac{c}{a}\right)$

Tangent

The tangent of an angle, in a right triangle, is equal to the side opposite the angle divided by the side adjacent to the angle.

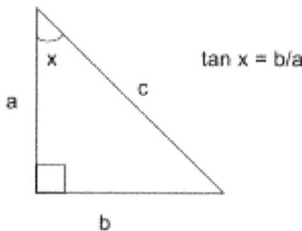

152. In the figure to the right, find tan A.

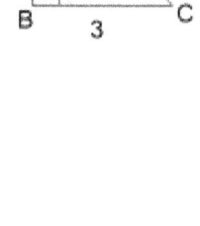

A. $\dfrac{7}{3}$

B. $\dfrac{3}{7}$

C. $\dfrac{3\sqrt{10}}{20}$

D. $\dfrac{2\sqrt{10}}{3}$

E. $\dfrac{2\sqrt{10}}{7}$

153. One of the angle measures is $\tan^{-1}(\dfrac{b}{a})$. What is $\sin(\tan^{-1}(\dfrac{b}{a}))$?

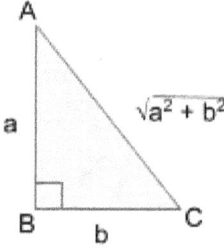

A. $\dfrac{b}{\sqrt{a^2+b^2}}$

B. $\dfrac{b}{a}$

C. $\sqrt{a^2+b^2}$

D. $\dfrac{a}{b}$

E. Unable to determine

> Hint: Re-draw and label, label, label to help you solve this problem.
> (Or write on your test booklet as needed.)

Trigonometric Identities

An identity is an equation that is always true. There are many trigonometric identities and relationships, but the ones commonly seen on the ACT are found in the Formulas section. Knowing how to apply these identities will help you in simplifying and solving expressions.

154. Simplify the expression $(\sin x + \cos x)^2 - 1$ using a Pythagorean Identity.

  A. 0
  B. $\sin^2 x$
  C. $2\sin x \cos x$
  D. $2\cos x$
  E. $2\sin x$

155. Rewrite the expression in terms of 1 function. $\dfrac{\sin^2 x}{\csc^3 x}$

  A. $\sin x$
  B. $\sin^2 x$
  C. $\sin^5 x$
  D. $\csc^5 x$
  E. Unable to determine

> Do not leave any questions blank. Guess if you have to. You are not penalized for incorrect answers.

Law of Sines

For the law of sines, we are given a relationship between the sine of the angles of a triangle and their respective side lengths. The relationship is given by the following:

$$\frac{\sin(A)}{a} = \frac{\sin(B)}{b} = \frac{\sin(C)}{c}$$

You may see the reciprocal of this relationship as well. Use the portion of the relationship that you need to solve your unknown(s).

Note that in the past, this relationship has either been stated in words in the question or given in formula form, but that is no guarantee that it will be given in future tests. Be familiar with this relationship and how to apply it.

156. In △PQR, what is the expression for the length of QR? (Note: The law of sines states that, for any triangle, the ratios of the sines of the interior angles to the lengths of the sides opposite those angles are equal.)

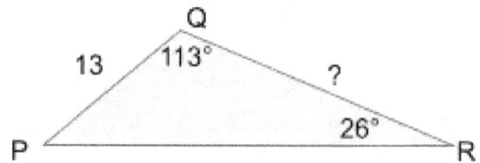

A. $\dfrac{13\sin(113)}{\sin(26)}$

B. $\dfrac{13\sin(26)}{\sin(41)}$

C. $\dfrac{13\sin(41)}{\sin(26)}$

D. $\dfrac{\sin(41)}{\sin(26)}$

E. $\dfrac{12\sin(41)}{\sin(26)}$

> Use any notes given to you in the question. That is usually a clue as to how the question should be worked.

157. In △ABC, AB is 20 units long, AC is 13 units long, and the measure of ∠B is 22°. What is the measure of ∠A, to the nearest tenth?

A. 2.0°
B. 35.19°
C. 55.0°
D. 122.8°
E. 125.3°

> You may see the Law of Sines in the following form:
> $$\dfrac{a}{\sin A} = \dfrac{b}{\sin B} = \dfrac{c}{\sin C}$$

Law of Cosines

For a non-right triangle, the law of cosines gives you a relationship between a side and the two other sides and the angle opposite the first side. For triangle ABC below, you can solve for any unknown side (or an unknown angle if given all 3 sides), given the previous conditions for the law of cosines.

$$c^2 = a^2 + b^2 - 2ab \cos C$$

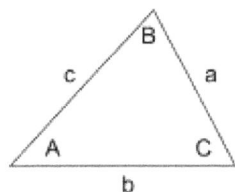

158. What is the measure of ∠A, to the nearest tenth?

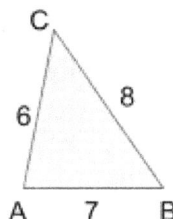

- A. 15.2
- B. 63.3
- C. 75.5
- D. 83.3
- E. 122.1

> If your unknown side is side a (or side b), then rearrange the Law of Cosines formula:
> $$a^2 = b^2 + c^2 - 2bc \cos A$$
> $$b^2 = a^2 + c^2 - 2ac \cos B$$

159. If a = 6, c = 9, and B = 128°, then what is b, to the nearest tenth?

- A. 10
- B. 11.3
- C. 13.5
- D. 20
- E. 183.5

Statistics and Probability

The Statistics & Probability subcategory accounts for between 8 and 12% of the total questions on the ACT.

We will work the following types of problems in this section:
- Fundamental Counting Principle
- Combinations
- Permutations
- Standard Deviation
- Probability
- Bivariate Data

Fundamental Counting Principle

In the Fundamental Counting Principle, if there are b ways to do one thing and c ways to do something else, then there are (b x c) ways to do both of these things.

160. In a florist, customers can choose from 5 types of roses, 3 types of greenery, and 4 colors of tissue paper. How many different bouquets are possible?

    A.   5
    B.   12
    C.   60
    D.   65
    E.   100

161. There are 5 students in a class. How many different ways can the 5 students form a line?

    A.   5
    B.   6
    C.   10
    D.   15
    E.   120

> Draw pictures or enumerate solutions as needed. Label figures so it is easier to remember and check later.

162. A customer must set a 4-digit PIN for his bank account using the number 0-9. How many possible outcomes are there for the PIN?

    A. 9^4
    B. 10^4
    C. 10^{10}
    D. 10^{14}
    E. 10^{40}

Combinations

When order of selection does not matter, you are dealing with a combination.

163. A committee will be selected from a group of 15 women and 12 men. The committee will be made up of 2 women and 2 men. Which of the following expressions gives the number of different committees that could be selected from these 27 people?

    A. $_{27}C_4$
    B. $(_{15}C_2)(_{12}C_2)$
    C. $(_{27}C_2)(_{27}C_2)$
    D. $(_{15}P_2)(_{12}P_2)$
    E. $_{27}P_4$

Permutations

Permutations should be calculated when order matters in a selection.

164. How many ways can we award a 1st place ribbon and a 2nd place ribbon among 20 contestants?

    A. $_{10}C_2$
    B. $_{20}C_2$
    C. $_{10}P_2$
    D. $_{18}P_2$
    E. $_{20}P_2$

> Hint: When order matters (1st or 2nd), use permutations. If we just award 2 people "superior" ribbons, the order does not matter...combinations.

Probability

Probability is the likelihood of some event occurring. It is a value between 0 and 1. Probability is similar to percents and fractions in that it is a part over a whole. Often, we talk about the number of times an event occurs divided by the number of all events occurring.

165. An integer from 1-100 is to be chosen randomly. What is the probability, to the nearest hundreth, that the number chosen will have 1 as at least 1 digit?

    A. 0.15
    B. 0.20
    C. 0.22
    D. 0.33
    E. 0.50

> Probability can be expressed as a percent, a decimal, or a fraction.

166. A security code is generated with the format of 1 numerical digit (0-9), 5 alphabetical letters, and 1 numerical digit (0-9). The numerical digits can repeat, but the alphabetical letters cannot repeat. What is the probability of generating the code with the letters "MUSIC"?

    A. $\dfrac{10^2}{(10^2)(26)(25)(24)(23)(22)}$

    B. $\dfrac{1}{10^2}$

    C. $\dfrac{1}{26^2}$

    D. $\dfrac{10^2}{(10^2)(26)}$

    E. $\dfrac{1}{(10^2)(26)(25)(24)(23)(22)}$

Standard Deviation

Standard deviation refers to how spread out the numbers in a data set are. Use the formula to calculate or visually inspect the data if possible.

167. If Matt's teacher decides to only record 4 grades for the grading period, what is the standard deviation of Matt's 4 test scores: 92, 85, 96, and 100?

    A. 1.39
    B. 5.54
    C. 11.08
    D. 93.25
    E. 122.75

> For standard deviation of a population, (1) find the mean, (2) square the difference of the mean and each data point, (3) find the mean of the squared values, (4) take the square root.

168. Sarah has 12 rose bushes, but decides to record the number of blooms from only 4 of these plants over a certain time period. When she has all the data she plans to collect, she decides to calculate the standard deviation of the number of blooms per rose bush. What number will she divide the sum of squared differences by?

    A. 3
    B. 4
    C. 12
    D. 20
    E. 48

> The sample standard deviation (s) is calculated when you only have data for a subset of the population. Thus, the sum of squared differences is divided by N-1, where N is the sample size.

169. Which set of numbers has a smaller standard deviation?

    A. 8, 12, 14, 3, 1, 6, 9
    B. 5, 1, 7, 9, 4, 12, 2
    C. 10, 5, 1, 12, 11, 2, 1
    D. 4, 12, 8, 9, 1, 10, 3
    E. 3, 4, 6, 2, 3, 5, 4

> The standard deviation can be thought of as the spread of the numbers within a data set. Visually inspect the numbers in this example and determine which set is spread out the least (in this question).

Bivariate Data

Bivariate data has to do with collecting data for 2 variables and comparing the data to determine any relationships.

170. A softball concessions director wants to determine if temperature affects snow cone sales to better gauge how many supplies to order for summer tournaments. If the statistical software she uses calculates a correlation coefficient (r) of 0.97, which scatter plot below could represent the data she collected?

A.

C.
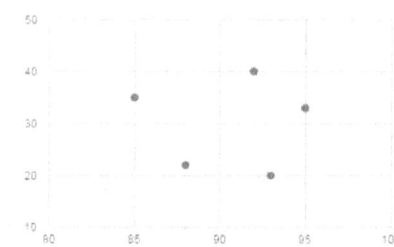

B.
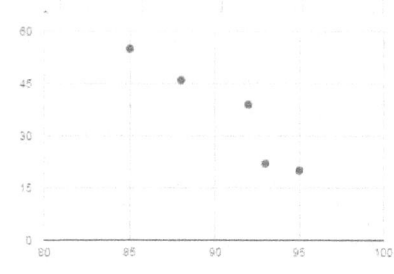

> The correlation coefficient, r, is used to give a measure as to how well a linear model describes a relationship between 2 variables.

171. Which scatter plot could represent data with r = 0?

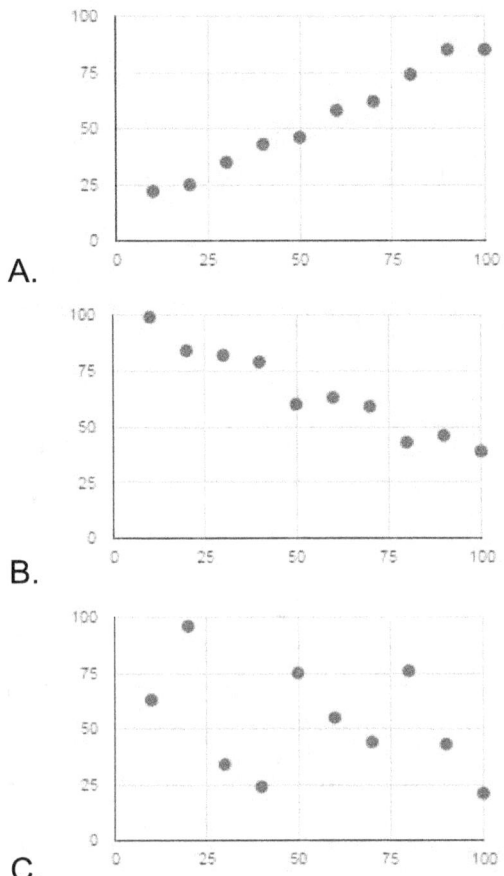

A.

B.

C.

Be prepared, do your best, and stay calm!

Integrating Essential Skills

Average/Mean, Median, and Mode

1. Matt has the following grades in his science class: 92, 85, 96, and 100. If he has one more test to take in the class, what must he make to finish the grading period with an average of 90?

   A. 77
   B. 79
   C. 82
   D. 90
   E. 93

   $$\begin{array}{r} 92 \\ 85 \\ 96 \\ 100 \\ \hline 373 \end{array}$$ already have $+ 1$ more test $\Rightarrow 5$ grades

   $\dfrac{x}{5} = 90$ (avg we want)

   $x = 90(5) = 450 \leftarrow$ goal total

   $450 - 373 = 77 \leftarrow$ min pts. needed

2. The baseball team sold meal tickets to raise money for the season. If the following number of tickets were sold by each of the players, what is the mean, median, and mode for ticket sales: 20, 15, 32, 22, 25, 20, 24, 27, 20, 18?

   A. 20.3, 21, 20
   B. 25.3, 21, 18
   C. 22.3, 21, 20
   D. 27.1, 21, 20
   E. 27.1, 25.3, 21

   Mean: $\dfrac{20+15+32+22+25+20+24+27+20+18}{10} = \dfrac{223}{10} = 22.3$

   *Note: If you feel 100% confident in your answer here, you could stop b/c choice C is the only one with this mean.

   Median: 10 numbers → avg of 5th & 6th values

   15 18 20 20 20 22 24 25 27 32
                        ↓
                       (21)

   Mode: occurs most often
   3 - 20's

SOLUTIONS - teachmathwithme.com   83

3. Susan is an advertising manager and prepares a monthly report for her supervisor. Susan's monthly sales are in the chart below. What is the average number of Susan's ads per magazine, to the nearest 0.1?

Number of Susan's Ads in a Magazine	Number of Magazines with this Total
1	3
2	5
3	7
4	4
5	2

A. 2.9
B. 3.2
C. 5.1
D. 7.0
E. 8.4

$\frac{60}{21} \approx 2.9$

Factors

4. What are all the positive factors of 10?

    A. 1, 10
    B. 1, 2, 5, 10
    C. 2, 10
    D. 10, 20, 30
    E. 5, 10, 15

    10: 1, 2, 5, 10

5. Which of the following lists all the positive factors of 9?

    A. 1, 9
    B. 3
    C. 3, 6
    D. 1, 3, 9
    E. 9, 18, 27

    9: 1, 3, 9

Least Common Multiple

6. What is the least common multiple of 5, 10, and 25?

    A. 2
    B. 10
    C. 25
    D. 50
    E. 100

    5 — 5, 10, ... , 45, (50) ...
    10 — 10, 20, 30, 40, (50) ...
    25 — 25, (50), ...

7. The local deli serves pasta salad every 4 days and pretzel salad every 6 days. If pasta salad and pretzel salad are both on today's menu, how many days before they are both on the menu again?

    A. 1
    B. 4
    C. 6
    D. 8
    E. 12

    PaS: 4, 8, (12), 16, ...
    PretS: 6, (12), ...

Fractions

8. Put the following fractions in increasing order: $\frac{1}{2}, \frac{3}{8}, \frac{1}{4}, \frac{5}{16}, \frac{7}{8}$

    A. $\frac{1}{4}, \frac{5}{16}, \frac{3}{8}, \frac{1}{2}, \frac{7}{8}$

    B. $\frac{1}{2}, \frac{5}{16}, \frac{7}{8}, \frac{1}{4}, \frac{3}{8}$

    C. $\frac{7}{8}, \frac{1}{2}, \frac{3}{8}, \frac{5}{16}, \frac{1}{4}$

    D. $\frac{1}{4}, \frac{1}{2}, \frac{3}{8}, \frac{5}{16}, \frac{7}{8}$

    E. $\frac{1}{4}, \frac{1}{2}, \frac{3}{8}, \frac{7}{8}, \frac{5}{16}$

    1 method — get common deno.

    $\frac{1}{2} = \frac{8}{16}$ (4)
    $\frac{3}{8} = \frac{6}{16}$ (3)
    $\frac{1}{4} = \frac{4}{16}$ (1)
    $\frac{5}{16}$ (2)
    $\frac{7}{8} = \frac{14}{16}$ (5)

SOLUTIONS - TeachMathWithMe.com

9. When 6 ¾ is written as an improper fraction, in lowest terms, what is the numerator?

   A. 3
   B. 18
   C. 24
   D. 27
   E. 30

$$6\frac{3}{4} = \frac{(4 \times 6) + 3}{4} \quad \frac{(Num)}{Deno}$$

$$= \frac{27}{4}$$

10. What is the least common denominator for adding $\frac{1}{6}$, $\frac{3}{8}$, $\frac{5}{12}$?

    A. 6
    B. 12
    C. 24
    D. 48
    E. 576

→ $\frac{1}{6}$ 6: 6, 12, 18, 24, ...

→ $\frac{3}{8}$ 8: 8, 16, 24, ...

→ $\frac{5}{12}$ 12: 12, 24, ...

Reciprocals

11. If 2 reciprocals are multiplied together, what is their product?

    A. 0
    B. 1
    C. 2
    D. 5
    E. 10

Ex. $\frac{1}{2}$ and 2 are reciprocals

$$\frac{1}{2}\left(\frac{2}{1}\right) = \frac{2}{2} = 1$$

12. What is the reciprocal of ⅝ ?

    A. $\frac{1}{8}$
    B. $\frac{5}{8}$
    C. $\frac{8}{5}$
    D. 5
    E. 8

$$\frac{5}{8} \Rightarrow \frac{8}{5}$$

Percent

13. If a $350 laptop is discounted 15%, but then has 9% sales tax added, what is the total price you must pay?

   A. $31.50
   B. $297.50
   C. $324.28
   D. $381.50
   E. $402.50

$$350(.15) = 52.50 \Rightarrow 350 - 52.50 = 297.5$$

$$297.5(.09) = 26.78 \Rightarrow 297.50 + 26.78 = 324.28$$

Use the following table for questions 14 and 15.

Menu Item	Number of Favorite Votes
Pizza	19
Hamburger	9
Salad	5
Chicken Tenders	14
Sloppy Joes	3

14. A poll of 50 random students in a high school was taken concerning a favorite lunch menu item. The results are in the table above. What percent voted for chicken tenders?

   A. 14%
   B. 20%
   C. 25%
   D. 28%
   E. 30%

$$\frac{14}{50} = 0.28 \Rightarrow 28\%$$

15. If this poll is indicative of how the 800 students in the entire school would vote, what is the best estimate of the number of votes hamburger would receive from the entire school?

   A. 9
   B. 50
   C. 144
   D. 150
   E. 450

$$\frac{9}{50} = \frac{x}{800}$$

$$50x = 7200$$
$$x = 144$$

SOLUTIONS - TeachMathWithMe.com

16. As a jewelry consultant, Sue makes a commission off each item she sells. If her sales are $500, she makes a commission of $150. How much commission does she earn if she makes $850 in sales?

    A. $150
    B. $ 255 ← (circled)
    C. $ 300
    D. $ 500
    E. $ 850

    $$\frac{150}{500} = \frac{x}{850}$$
    $$500x = (850)(150) = 127,500$$
    $$x = \$255$$

17. Julie visited a craft store where all items were <u>15% off</u>. She wants to program her calculator so she can input the marked price and the discounted price will be output. What is the expression for the discounted price on a marked price of p dollars?

    A. 0.15p — discount only
    B. 0.15p + p — added the discount to price
    C. 0.15p + 0.85p — total price w/o discount
    D. p - 0.15p ✱ ← (circled)
    E. 0.15p - p — discount minus price

18. 250 students entered an art contest. The entries are divided into 4 categories as shown below. There are 50 prizes to be awarded. If the prizes are to be awarded in proportion to the number of entries in each category, how many prizes should be in the 10-11 year old category?

Age Category	Under 9	10-11	12-13	14-15
Number of Entries	70	80	65	35

    A. 13
    B. 16 ← (circled)
    C. 20
    D. 25
    E. 80

    $$\frac{80}{250} = \frac{x}{50}$$
    $$250x = 4000$$
    $$x = 16$$

SOLUTIONS - teachmathwithme.com

Ratios

19. A group of 3 friends share and eat a whole pizza. Carrie eats 1 piece, John eats 4 pieces, and Sam eats 3 pieces. What is the ratio of Carrie's share to John's share to Sam's share?

    A. 3:4:1
    B. 4:3:1
    C. 7:1
    D. 1:8
    E. 1:4:3 ← (circled)

    C : J : S
    1 : 4 : 3

20. Jane fills a bowl with colored jelly beans. There are 20 red, 16 yellow, 10 green, and 14 purple. In lowest terms, what is the ratio of red to yellow to green to purple jelly beans?

    A. 1:4:6:2
    B. 10:8:5:7 ← (circled)
    C. 8:5:7:10
    D. 5:4:2:3
    E. 10:16:1:14

    R : Y : G : P
    20 : 16 : 10 : 14   } lowest ÷ 2
    10 : 8 : 5 : 7   terms

Rational Numbers

21. Which of the choices is NOT a rational number?

    A. $\sqrt{\frac{5}{25}}$ → unending decimal value ← (circled)
    B. $\frac{1}{3}$ ✓ fraction
    C. $\frac{1}{2}$ ✓ fraction
    D. $5 = \frac{5}{1}$ ✓ fraction
    E. .21212121... ✓ repeating decimal → fraction

    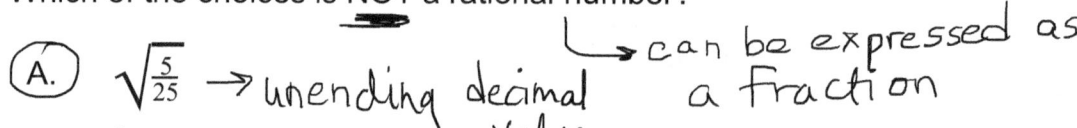
    → can be expressed as a fraction

SOLUTIONS - teachmathwithme.com

22. What rational number is halfway between $\frac{1}{2}$ and $\frac{1}{3}$?

    A. $\frac{1}{4}$

    B. $\frac{1}{5}$

    (C.) $\frac{5}{12}$

    D. $\frac{3}{8}$

    E. $\frac{7}{12}$

$$\frac{1}{2} + \frac{1}{3} = \frac{3}{6} + \frac{2}{6} = \frac{5}{6}$$

$$\frac{5/6}{2} = \frac{5}{6} \times \frac{1}{2} = \frac{5}{12}$$

Scientific Notation

23. What is 0.00000325 in scientific notation?

    A. 3.25
    B. 325
    (C.) 3.25×10^{-6}
    D. 3.25×10^{6}
    E. 3250

$$3.25 \times 10^{-6}$$

24. What is 438000000000 in scientific notation?

    A. 4.38×10^{-11}
    B. 4.38×10^{-4}
    C. 4.38
    D. 438
    (E.) 4.38×10^{11}

$$4.38 \times 10^{11}$$

Data Interpretation

25. Jenn is planning to buy cookies from a bakery for her cousin's birthday party. She needs 50 cookies. When she gets to the bakery, she sees the chart below. What's the minimum price Jenn will pay for 50 cookies?

Individual	1 Dozen	2 Dozen
$0.50	$5.50	$10.75

2 × 12 = 24

A. $10.75
B. $21.50
C. $22.50 ← circled
D. $23.00
E. $25.00

2 - 2 Dozens = 48 → 10.75 × 2 = 21.50

2 - Individuals = 2

50 ✓ → 7.5 × 2 = 1.00

Rates

21.50 + 1.00 = $22.50

26. Mike is driving to see the New York Yankees play. It is 300 miles to the interstate exit that he needs to reach. He drives an average speed of 65 miles/hour. If he has already driven 3 hours, but wants to get to the exit in 1.5 more hours, how many miles per hour faster should he drive?

A. 1.5 miles/hr
B. 5 miles/hr ← circled
C. 50 miles/hr
D. 65 miles/hr
E. 70 miles/hr

65 miles/hr (3 hr) = 195 miles ahead

300 - 195 = 105 miles to go

105 miles / 1.5 hr = 70 m/h

70 - 65 = 5 m/h faster

27. A machine makes 10 widgets per minute. A second machine makes 25 widgets per minute. The second machine starts making widgets 3 minutes after the first machine starts. Both machines stop making widgets 10 minutes after the first machine started. Together, the 2 machines make how many widgets?

A. 35
B. 100
C. 200
D. 275 ← circled
E. 350

m/c 1: 10 w/min (10 min) = 100 widgets

m/c 2: 25 w/min (10 - 3 min) = 175 widgets

275

SOLUTIONS - TeachMathWithMe.com

Perimeter

28. Figure ABCD is a square with side length of 4. Point E is the midpoint of AB, point F is the midpoint of BC, and point G is the midpoint of CD. What is the perimeter of the region AEFGD?

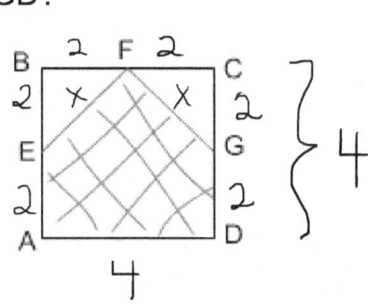
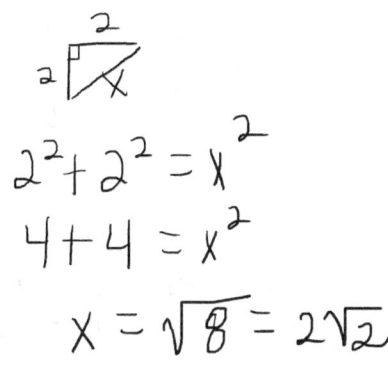

A. 8
B. $4 + 2\sqrt{2}$
C. $8 + 4\sqrt{2}$ ✓
D. 16
E. $12 + 2\sqrt{2}$

$2^2 + 2^2 = x^2$
$4 + 4 = x^2$
$x = \sqrt{8} = 2\sqrt{2}$

$P = 4 + 2 + 2 + 2\sqrt{2} + 2\sqrt{2}$
$= 8 + 4\sqrt{2}$

29. The perimeter of a parallelogram is 32 ft. The length of one side is 9 ft. What are the lengths, in ft., of the other 3 sides?

A. 4, 4, 4
B. 9, 4, 4
C. 9, 7, 7 ✓
D. 8, 8, 8
E. Cannot be determined

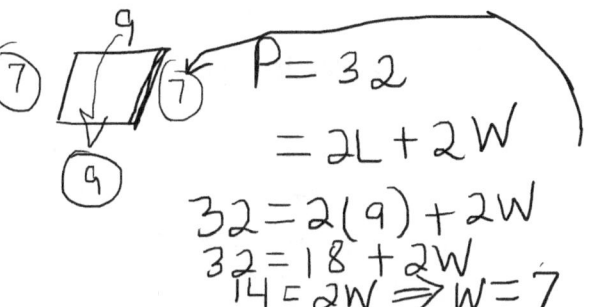

$P = 32 = 2L + 2W$
$32 = 2(9) + 2W$
$32 = 18 + 2W$
$14 = 2W \Rightarrow W = 7$

30. In the figure below, adjacent sides meet at right angles and lengths given are in feet. What is the perimeter of the figure, in feet?

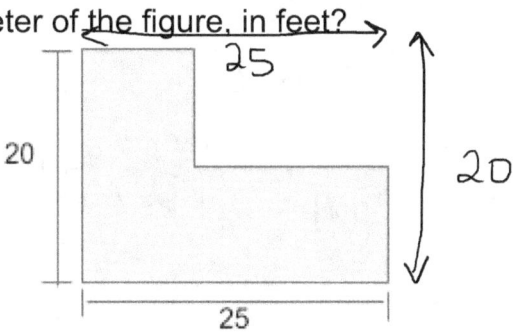

A. 20
B. 25
C. 45
D. 90 ✓
E. Cannot be determined

$25(2) + 20(2)$
$50 + 40 = 90$

SOLUTIONS - TeachMathWithMe.com   92

Circumference

31. What is the circumference of a circle with radius of 3 units? Use 3.14 for π.

    A. 4.71
    B. 5.32
    C. 9.42
    D. 18.84 ✓
    E. 56.52

    $C = 2\pi r$
    $= 2(3.14)(3) = 18.84$

32. An 8.5" diameter bowling ball is released at the foul line and starts rolling towards the head pin. If it is 60 feet from the foul line to the head pin, how many revolutions will the ball make before it hits the head pin?

    A. 12.2
    B. 13.35
    C. 26.69
    D. 26.98 ✓
    E. 30

    $\dfrac{60 ft}{2.225 ft} \approx 26.97$

    Revolutions → Circumference
    $C = \pi d = (8.5\pi)$ in $\times \dfrac{1 ft}{12 in} \approx 2.225 ft$

Area - Rectangle

33. Mike plans to use grass seed in the backyard of his newly built home. His backyard is a 36 ft x 40 ft rectangular plot, but he has installed a 12 ft x 24 ft inground pool. If each bag of grass seed covers 375 ft², how many bags will he need to purchase?

    A. 2
    B. 3
    C. 4 ✓
    D. 5
    E. 6

    $A_{Backyard} = 36 \times 40 = 1440 ft^2$

    $A_{pool} = 12 \times 24 = \underline{288 ft^2}$

    $1152 ft^2$ of grass

    $\dfrac{1152}{375} = 3.072$ bags ⇒ 4

34. What percent of the area of the larger rectangle is the area of the smaller, shaded rectangle?

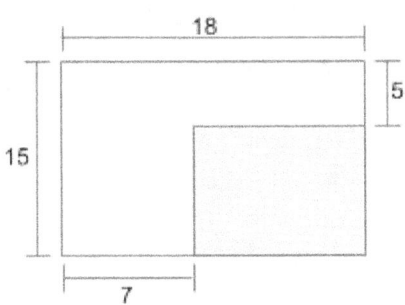

A. 25.3%
B. 32.7%
C. 40.7%
D. 42.7%
E. 57.6%

$A_L = 15 \times 18 = 270$

$A_S = (18-7)(15-5) = 11 \times 10 = 110$

$\frac{110}{270} \approx .407 \Rightarrow 40.7\%$

35. A rectangle has an area of 36 ft² and a perimeter of 26 feet. What is the longest of the side lengths, in feet, of the rectangle?

A. 4
B. 5.5
C. 6
D. 9
E. 10

$A = L \times W = 36 \Rightarrow (13-W)W = 36$
$13W - W^2 = 36 \Rightarrow W^2 - 13W + 36 = 0$
$(W-9)(W-4)$
$W = 9, W = 4$

$P = 2L + 2W = 26 \Rightarrow 2L = 26 - 2W$
$L = 13 - W$
$L = 9, 4$

36. In the figure below, the vertices of rectangle ABCD have (x,y) coordinates. What is the area of rectangle ABCD?

A. 10
B. 18
C. 20
D. 22
E. 30

$(7-2) = 5$
$6-2 = 4$

$A = L \times W = 5 \times 4 = 20$

Area - Circle

37. What is the area, in m², of a circle with diameter 6 m? Use 3.14 for π.

    A. 9.42
    B. 12.52
    C. 18.84
    D. 28.26
    E. 113.04

    $r = 3$
    $A = \pi r^2 = (3.14)(3)^2 = 28.26$

38. Jill installed a 9 ft. leash to a stake right outside her front door. What is the outside area, in square ft., the dog can reach from the stake?

    A. 4.5 π
    B. 9 π
    C. 40.5 π
    D. 81 π
    E. 105 π

    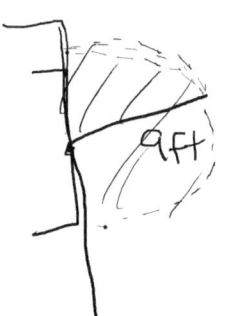

    ½ Area of circle w/ r = 9
    $A = \pi r^2 = 81\pi$
    $\frac{1}{2}(81\pi) = 40.5\pi$

Area - Triangle

39. What is the area of the right triangle with side lengths 5 cm and 7 cm?

    A. 12.0
    B. 17.5
    C. 19.3
    D. 35.0
    E. 36.2

    $A = \frac{1}{2}bh = \frac{1}{2}(7)(5) = 17.5$

40. Given the figure below, what is the area of the equilateral triangle ABC?

    A. 5
    B. 25 √3
    C. 30
    D. 75
    E. 100

    $h^2 + 5^2 = 10^2$
    $h^2 = 100 - 25 = 75$
    $h = 5\sqrt{3}$
    $A = \frac{1}{2}bh = \frac{1}{2}(10)(5\sqrt{3}) = 25\sqrt{3}$

41. In the figure below, the vertices of △ABC have (x,y) coordinates. What is the area of △ABC?

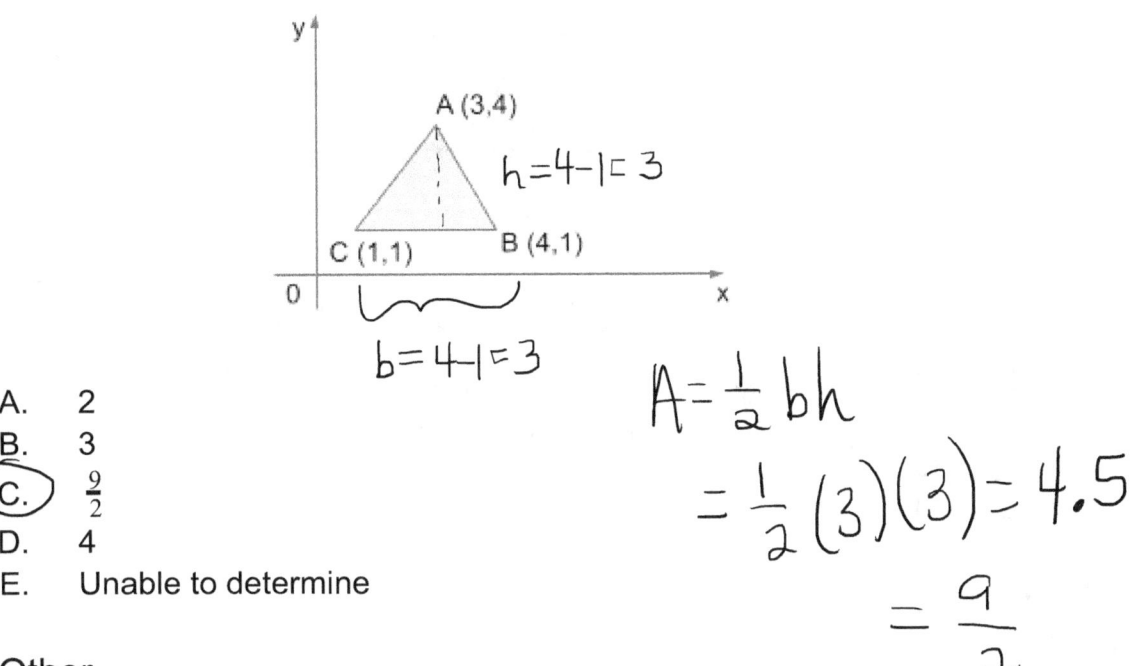

A. 2
B. 3
C. $\frac{9}{2}$ ← circled
D. 4
E. Unable to determine

$h = 4-1 = 3$
$b = 4-1 = 3$

$A = \frac{1}{2}bh$
$= \frac{1}{2}(3)(3) = 4.5$
$= \frac{9}{2}$

Area - Other

42. In the isosceles trapezoid given below, the parallel lines are 8 in and 14 in long, respectively. What is the area of the trapezoid, in in²?

8 in.
5 in. / h \ 5
3 14 in. 3

A. 24
B. 32
C. 44 ← circled
D. 48
E. 52

$h^2 + 3^2 = 5^2$
$h^2 = 25 - 9$
$h^2 = 16$
$h = 4$

$A = \frac{1}{2}h(b_1 + b_2)$
$= \frac{1}{2}(4)(14+8)$
$= 44$

43. Given the measurements in the figure below and that ∠A = ∠D and BC is parallel to FE, what is the area, in square units, of the hexagon?

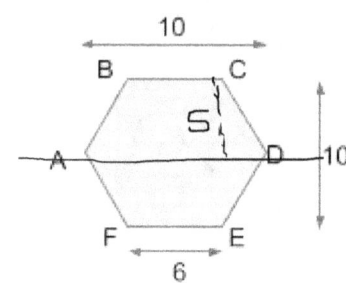

A. 20
B. 60
C. 70
D. 80
E. 90

1 method — split into 2 trapezoids

$$A = \tfrac{1}{2} h (b_1 + b_2) = \tfrac{1}{2} \cdot 5(10+6) = \tfrac{1}{2} \cdot 5 \cdot \overset{8}{\cancel{16}}$$
$$= 40$$

$$2(40) = 80$$

Midpoint

44. What is the midpoint of the line with endpoints (1,2) and (7, -2)?

A. (0,1)
B. (1,0)
C. (3, 2)
D. (4,0)
E. (4,1)

$$\text{mid} = \left(\frac{x_1 + x_2}{2}, \frac{y_1 + y_2}{2} \right)$$
$$= \left(\frac{1+7}{2}, \frac{2+-2}{2} \right) = (4, 0)$$

45. The number line below is the graph of which of the following inequalities?

≤ 0   ≥ 2

$0 \geq x \geq 2$

A. $0 \geq x \geq 2$
B. $0 \leq x \leq 2$
C. $x \geq 2$
D. $x \leq 0$
E. $-1 < x < 2$

Preparing for Higher Math

Number & Quantity
Absolute Value

46. Evaluate the expression: | (3)(-2) - (6)(4) | - | (2)(-1) + (3)(5) |.

    A. -43
    B. -17
    C. 10
    D. 17 ⭕
    A. 43

    $|-6 - 24| - |-2 + 15|$
    $|-30| - |13|$
    $30 - 13 = 17$

47. -3 | 7 - 5 | + 2 | -3 - 2 | = ?

    A. -72
    B. -16
    C. 1
    D. 4 ⭕
    E. 7

    $-3|2| + 2|-5|$
    $-3(2) + 2(5) = -6 + 10 = 4$

Multiplying and Dividing Variables

48. What is the product of $(3xy^2)(-2x^3y^3)$?

    A. $-6x^3y^6$
    B. $-6x^4y^5$ ⭕
    C. $3x^3y^6$
    D. $6x^4y^5$
    E. $6x^3y^6$

    $-6x^{1+3}y^{2+3}$
    $-6x^4y^5$

49. Evaluate the expression $\dfrac{x^3y^2z}{x^{-2}z^3}$.

    A. $\dfrac{x^5y^2}{z^2}$ ⭕
    B. $\dfrac{x^6y^2}{z^2}$
    C. x^5y^2
    D. $\dfrac{x^6y^2}{z^3}$
    E. $\dfrac{x^5y^2}{z^3}$

    $x^{3-(-2)}y^2z^{1-3} = x^5y^2z^{-2}$
    $\dfrac{x^5y^2}{z^2}$

Exponents

50. What is the following expression equivalent to? $\dfrac{(3x^2y)^3}{xy^2}$

   A. 8xy
   B. $9x^5y$
   C. $\dfrac{3x^5y}{xy}$
   D. $27x^5y$ ◉
   E. xy

*could really stop right here b/c D is only answer w/ 27.

$$\dfrac{3^3 x^{2\times 3} y^{1\times 3}}{xy^2} = \dfrac{27x^6 y^3}{xy^2}$$
$$= 27 x^{6-1} y^{3-2}$$
$$= 27 x^5 y$$

51. What is $(2a^2b)^3$?

   A. $2a^6b^3$
   B. $2a^5b^4$
   C. $4a^5b^4$
   D. $8a^6b^3$ ◉
   E. $10a^6b^3$

$2^3 a^{2\times 3} b^{1\times 3}$

$8 a^6 b^3$

*Again, could stop when you get 8 & your answer choices.

Imaginary Numbers

52. For $i^2 = -1$, $(2+i)^2 = ?$

   A. 3
   B. 4 + 4i
   C. 2 + 3i
   D. 3 + 4i ◉
   E. 4i

$(2+i)(2+i)$

$4 + 2i + 2i + i^2$
$4 + 4i + (-1)$
$3 + 4i$

Matrices

53. Multiply the following matrices.

$\begin{pmatrix} 3 & 2 & 1 \\ 0 & 4 & 2 \end{pmatrix} \begin{bmatrix} 1 \\ 2 \\ 3 \end{bmatrix} = \begin{bmatrix} 3(1)+2(2)+1(3) \\ 0(1)+4(2)+2(3) \end{bmatrix}$

- (A.) $\begin{bmatrix} 10 \\ 14 \end{bmatrix}$
- B. $\begin{bmatrix} 7 & 10 \end{bmatrix}$
- C. $\begin{bmatrix} 10 & 14 \end{bmatrix}$
- D. $\begin{bmatrix} 3 \\ 7 \end{bmatrix}$
- E. $\begin{bmatrix} 10 \\ 7 \end{bmatrix}$

$= \begin{bmatrix} 3+4+3 \\ 0+8+6 \end{bmatrix}$

$= \begin{bmatrix} 10 \\ 14 \end{bmatrix}$

54. Find the determinant of the matrix.

$\begin{bmatrix} 3 & 2 \\ 4 & 1 \end{bmatrix}$

- A. -7
- (B.) -5
- C. 2
- D. 5
- E. 9

$\det = (3)(1) - 2(4)$
$= 3 - 8 = -5$

55. What is the resulting matrix for ?

$$4 \begin{pmatrix} 1 & 0 \\ 2 & 3 \end{pmatrix} + 2 \begin{pmatrix} 2 & 3 \\ 1 & 0 \end{pmatrix}$$

A. $\begin{pmatrix} 6 & 2 \\ 4 & 2 \end{pmatrix}$

B. $\begin{pmatrix} 8 & 6 \\ 10 & 12 \end{pmatrix}$ ✓

C. $\begin{pmatrix} 4 & 0 \\ 8 & 12 \end{pmatrix}$

D. $\begin{pmatrix} 3 & 2 \\ 4 & 1 \end{pmatrix}$

E. $\begin{pmatrix} 4 & 6 \\ 2 & 0 \end{pmatrix}$

$\begin{bmatrix} 4 & 0 \\ 8 & 12 \end{bmatrix} + \begin{bmatrix} 4 & 6 \\ 2 & 0 \end{bmatrix}$

$\begin{bmatrix} 4+4 & 0+6 \\ 8+2 & 12+0 \end{bmatrix}$

$\begin{bmatrix} 8 & 6 \\ 10 & 12 \end{bmatrix}$

Complex Numbers

56. What is the distance between the points (1 + 4i) and (3 - 2i) on the complex plane?

    A. $\sqrt{13}$
    B. 5
    C. $\sqrt{17}$
    D. $2\sqrt{10}$ ✓
    E. 15

$z = (1+4i) - (3-2i) = -2 + 6i$

$dist = \sqrt{(-2)^2 + 6^2} = \sqrt{4+36}$

$= \sqrt{40} = 2\sqrt{10}$

57. What is the midpoint between the points (1 + 4i) and (3 - 2i) on the complex plane?

    A. 2 - 3i
    B. 1 - i
    C. 2 + i ✓
    D. 4 + 2i
    E. 3 + 4i

$\left(\dfrac{1+3}{2} + \dfrac{4+(-2)}{2} i \right)$

$(2 + i)$

Vectors

58. What is the result of adding the vectors a = (3, 6) and b = (2, -5)?

   A. (1, 11)
   B. (5, 11)
   C. (5, 2)
   D. (5, 1) ✓
   E. (6, 3)

$$a+b = (3+2, 6+(-5))$$
$$= (5, 1)$$

Algebra

Evaluating Expressions

59. If x = 3 and y = 4, what is the value of z in the following expression: z = 2x + 3y?

   A. 7
   B. 9
   C. 18 ✓
   D. 21
   A. 30

$$z = 2(3) + 3(4)$$
$$= 6 + 12 = 18$$

60. The ideal gas law is the equation of state of a hypothetical ideal gas. The ideal gas law is often written as PV=nRT. Solve this equation for T, the temperature of the gas.

   A. T = PVnR
   B. T = $\frac{PV}{nR}$ ✓
   C. T = $\frac{nR}{PV}$
   D. T = $\frac{PnR}{V}$
   E. T = $\frac{P}{VnR}$

$$\frac{PV}{nR} = \frac{nRT}{nR}$$

$$T = \frac{PV}{nR}$$

61. Which equation represents a in terms of b for 2a - 8b = 4?

   A. 8b + 4
   B. 2b
   C. -4b - 2
   D. 4b + 2 ✓
   E. $\frac{b}{2}$

$$2a - 8b = 4$$
$$2a = 8b + 4$$
$$a = 4b + 2$$

62. If $(x,y) \bullet (a,b) = xb - ya$, what is $(1,2) \bullet (3,4)$?

    A. -2 ✓
    B. 2
    C. 3
    D. 4
    E. 6

$(x,y) \bullet (a,b) = xb - ya$

$(1,2) \bullet (3,4) = 1(4) - 2(3) = 4 - 6 = -2$

63. The formula for exponential decay is given by the following: $y = A(1-r)^t$, where y is the new value, A is the original value, r is rate as a decimal, and t is time in years. If the original value is 35, the rate is 12%, and the time is 5 years, what is y, to the nearest tenth?

    A. 13.2
    B. 15.9
    C. 18.5 ✓
    D. 22.7
    E. 30.1

$y = A(1-r)^t$

$y = 35(1 - .12)^5$

≈ 18.5

Writing Expressions

64. Write an expression for 3 more than 2 times a number.

    A. $3 + 2x$ ✓
    B. $2 + 3x$
    C. $2 + 3$
    D. $(2 + 3)x$
    E. $(2 + x)(3x)$

    $2x + 3$

65. Emilee designs websites for customers. Her fees include a $50 consultation fee plus a $30/hour design fee. How many hours did she charge on a $200 bill?

    A. 2 ⅓
    B. 4
    C. 5 ✓
    D. 6 ⅔
    E. 8

$50 + 30x = 200$

$30x = 150$

$x = 5$

66. The sum of two numbers is 12. Their difference is 6. What is their product?

   A. 9
   B. 15
   C. 27
   D. 30
   E. 42

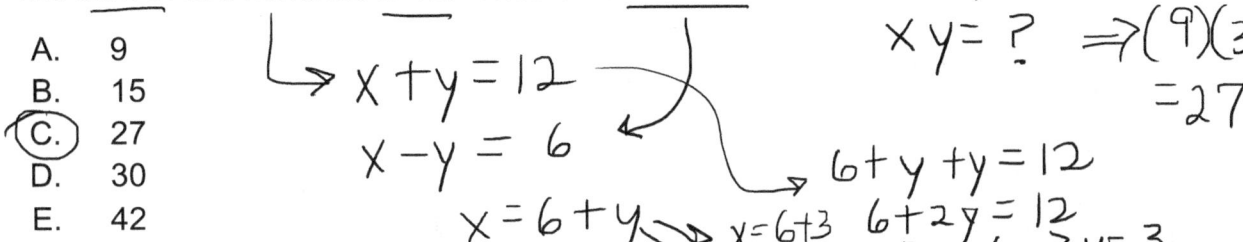

67. If x and y vary as given in the chart below, which of the following equations represents this relationship?

x	0	1	2	3	4
y	1	6	11	16	21

   A. $y = -5x + 1$
   B. $y = x + 5$
   C. $y = 5x + 1$
   D. $y = 5x + 6$
   E. $y = 21x$

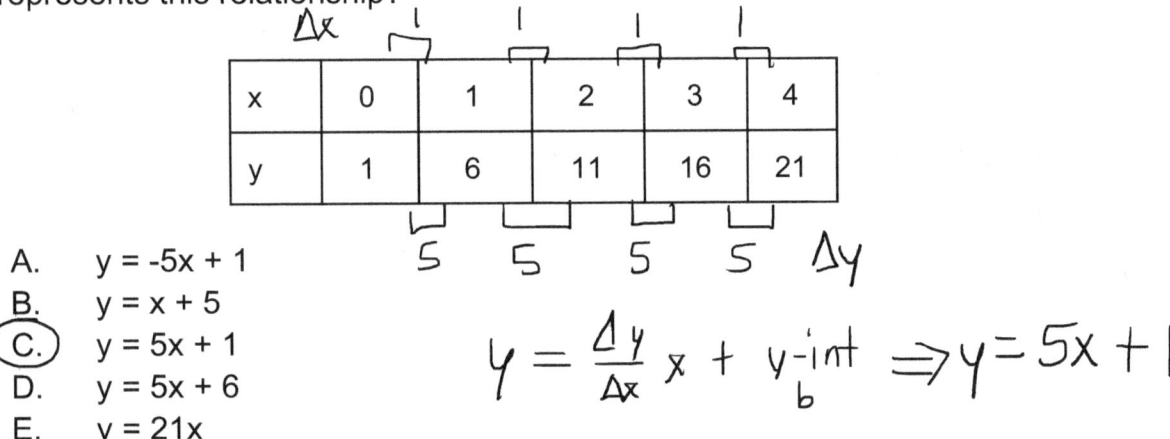

68. For 2 consecutive integers, the result of doubling the smaller integer and adding to the larger integer is 76. What are the 2 integers?

   A. 3 and 4
   B. 12 and 15
   C. 25 and 26
   D. 30 and 31
   E. 34 and 35

$x, x+1$

$2x + (x+1) = 76$
$3x = 75$
$x = 25 \Rightarrow x+1 = 26$

69. Regan is selling t-shirts for her summer business. She can rent a stand at the mall for $100/month. Each t-shirt costs her $2 to make, but she can sell one for $15. If she runs her business for 3 months at the mall, what is the expression that represents her profit when x t-shirts are produced and sold?

   A. 13x - 300
   B. 300 - 13x
   C. 15x - 100
   D. 13x
   E. 13x - 100

$100 \times 3 = 300$ (rent)

$(15 - 2) = \$13$ profit/shirt

$13x - 300$

70. If $2y = x^2 + 4$, then what is $-y$ equivalent to?

   A. $\frac{(-x)^2}{2} - 2$
   B. $\frac{-x^2}{2} - 2$ ✓
   C. $\frac{x^2}{2} - 2$
   D. $\frac{-x^2}{2} + 4$
   E. $\frac{-x^2}{2} - 4$

$2y = x^2 + 4$
$y = \frac{x^2}{2} + 2$
$\Rightarrow -y = -\frac{x^2}{2} - 2$

Combining Like Terms

71. What is $(3x + 2y - z) - (4x - 5y + 3z)$ equivalent to?

   A. $-x - 3y + 2z$
   B. $-x + 7y - 4z$ ✓
   C. $12x^2 + 10y^2 - 3z^2$
   D. $7x + 7y + 4z$
   A. $7x - 3y - 4z$

$(3x + 2y - z) - (4x - 5y + 3z)$
$-x + 7y - 4z$

72. What must be added to $x^2 + 5x$ so that the sum is $3x^2 - 2x + 4$?

   A. $3x - 2$
   B. $x^2 - 2x + 4$
   C. $2x + 4$
   D. $x^2 - 3x + 4$
   E. $2x^2 - 7x + 4$ ✓

$3x^2 - 2x + 4 - (x^2 + 5x)$
$2x^2 - 7x + 4$

Distributive Property

73. Evaluate the expression $-3xy(x^2 - 2xy^3)$.

   A. $-3x^3y - 6x^2y^4$
   B. $-3x^2y - 6x^2y^3$
   C. $-3x^3y + 6x^2y^4$ ✓
   D. $-3x^2y - 6x^2y^3$
   E. $-3x^3y$

$-3x^3y + 6x^2y^4$

Simplify Variables

74. Simplify the expression $(x - 1)/(x^2 - 2x + 1)$.

　A. $\frac{1}{x}$
　B. $\frac{1}{x-1}$ ✓
　C. $\frac{1}{x+1}$
　D. $\frac{x-1}{x+1}$
　E. 13

$$\frac{x-1}{(x-1)(x-1)} = \frac{1}{x-1}$$

75. For all positive real numbers x, which expression is equivalent to $\frac{\frac{x^{18}}{x^7}}{\frac{1}{x^3}}$?

　A. 1
　B. x^3
　C. x^8
　D. x^{14} ✓
　E. x^{33}

$$\frac{x^{18}}{x^7} \cdot \frac{x^3}{1} = \frac{x^{21}}{x^7} = x^{14}$$

FOIL

76. The expression $(x^2 + 6)(x - 3y)$ is equivalent to?

　A. $x^2 - 3xy + 6x - 18y$
　B. $x^3 - 18y$
　C. $x^2 - 3x^2y + 6x - 18y$
　D. $x^3 - 3x^2y + 6x - 18y$ ✓
　E. $x^3 - 3x^2y - 6x - 18y$

$(x^2 + 6)(x - 3y)$

$x^3 - 3x^2y + 6x - 18y$

77. What is $(z + 3x)^2$ equivalent to?

　A. $z^2 + 3$
　B. $z^2 + 9x^2$
　C. $z + 3x$
　D. $z^2 + 6xz + 9x^2$ ✓
　E. z

$(z + 3x)(z + 3x)$

$z^2 + 3xz + 3xz + 9x^2$

$z^2 + 6xz + 9x^2$

Factoring

78. What is a factored form of the expression $x^2 + x - 12$?
   - A. $(x - 4)(x - 3)$
   - B. $(x - 4)(x + 3)$
   - C. $(x + 4)(x - 3)$ ✓
   - D. $(x + 4)(x + 3)$
   - E. $(x + 6)(x - 2)$

$(x + 4)(x - 3)$

79. What values are the solution for $x^2 + x = 6$?
   - A. -3 and 2 ✓
   - B. 6 and 0
   - C. 3 and -2
   - D. 6 and 1
   - E. 1 and 0

$x^2 + x - 6 = 0$
$(x + 3)(x - 2) = 0$
$x = -3, x = 2$

Solving Equations

80. What is x equal to in the equation $3^{x+1} = 9^{2x+2}$?
   - A. -2
   - B. -1 ✓
   - C. 1
   - D. 2
   - E. 6

$3^{x+1} = 3^{2(2x+2)}$ get same bases
$x + 1 = 2(2x + 2)$
$x + 1 = 4x + 4$
$-3 = 3x$
$x = -1$

81. If $\frac{2}{\sqrt{5}} = \frac{a\sqrt{5}}{3}$, what is a?
   - A. $\sqrt{3}$
   - B. 2
   - C. $\frac{6}{5}$ ✓
   - D. 4
   - E. 5

$\left(\frac{3}{\sqrt{5}}\right)\frac{2}{\sqrt{5}} = \frac{a\sqrt{5}}{3} \times \frac{3}{\sqrt{5}}$
$\frac{6}{5} = a$

82. What are the real solutions to the following equation: $|x| - 3|x| + 10 = 0$?
   - A. 0, 5
   - B. -2, 5
   - C. -1, 4
   - D. -5, 5 ✓
   - E. 3, 10

$x - 3x + 10 = 0$
$-2x = -10$
$x = 5$ + b/c absolute value, x can be -5.

System of Equations

83. What is the value of x and y if given the following equations:

$$x + 3y = 6$$
$$2x + y = 3?$$

A. $x = 3$ and $y = \frac{9}{5}$
B. $x = \frac{3}{5}$ and $y = \frac{9}{5}$ ✓
C. $x = 3$ and $y = 9$
D. $x = \frac{2}{5}$ and $y = \frac{9}{5}$
E. $x = 5$ and $y = \frac{9}{5}$

$x + 3y = 6$
$\times (-3) \Rightarrow -6x - 3y = -9$

$-5x = -3$
$x = 3/5$

$\frac{3}{5} + 3y = 6$
$3y = \frac{27}{5}$
$y = 9/5$

84. For what value of *a* would the following system of equations have an infinite number of solutions?

$$(3x - y = 12) \, 4$$
$$12x - 4y = 8a$$

A. 3
B. 4
C. 6 ✓
D. 12
E. 48

$12x - 4y = 48$

$8a = 48$

$a = 6$

Slope

85. What is the slope of the line through (3,2) and (-4, 5) in the standard (x,y) coordinate plane?

A. $-\frac{1}{2}$
B. $-\frac{3}{7}$ ✓
C. 1
D. $\frac{3}{7}$
E. $\frac{1}{2}$

$m = \frac{\Delta y}{\Delta x} = \frac{5-2}{-4-3} = \frac{3}{-7}$

86. Given the equation 4x - 3y = 7, what is the slope of the line perpendicular to this line?

   A. $-\frac{7}{3}$
   B. $-\frac{3}{4}$
   C. $\frac{4}{3}$
   D. 3
   E. 4

$-3y = -4x + 7$
$y = \frac{4}{3}x - \frac{7}{3}$
$\hookrightarrow m_\perp = -\frac{3}{4}$

87. What is the slope of the line parallel to 8x - 3y = 1?

   A. -8
   B. $-\frac{3}{8}$
   C. $\frac{8}{3}$
   D. 3
   E. 8

$-3y = -8x + 1$
$y = \frac{8}{3}x - \frac{1}{3}$
$\hookrightarrow m_{//} = \frac{8}{3}$

88. What is the slope of the line perpendicular to 2y - 4x = 6?

   A. -4
   B. -2
   C. -½
   D. 2
   E. 4

$2y = 4x + 6$
$y = 2x + 3$
$\hookrightarrow m_\perp = -\frac{1}{2}$

Equation of Line

89. The graph of y = 2x + 4 passes through (8a, 12) in the standard (x,y) coordinate plane. What is the value of a?

   A. -2
   B. $\frac{1}{2}$
   C. $\frac{3}{2}$
   D. 2
   E. 4

$12 = 2(8a) + 4$
$12 = 16a + 4$
$8 = 16a$
$a = \frac{1}{2}$

90. Line t in the standard (x,y) coordinate plane has equation x = 4 and intersects line u given by equation y = x + 1. What is the point of intersection of lines t and u?

   A. 0
   B. 1
   C. 4
   (D.) 5
   E. 7

$$y = 4 + 1$$
$$y = 5$$

Graph of Line

91. The point (1,3) is shown in the standard (x,y) coordinate plane. If there is a line that passes through (1,3) with slope ⅓, what is another point on that line? Use the plane below.

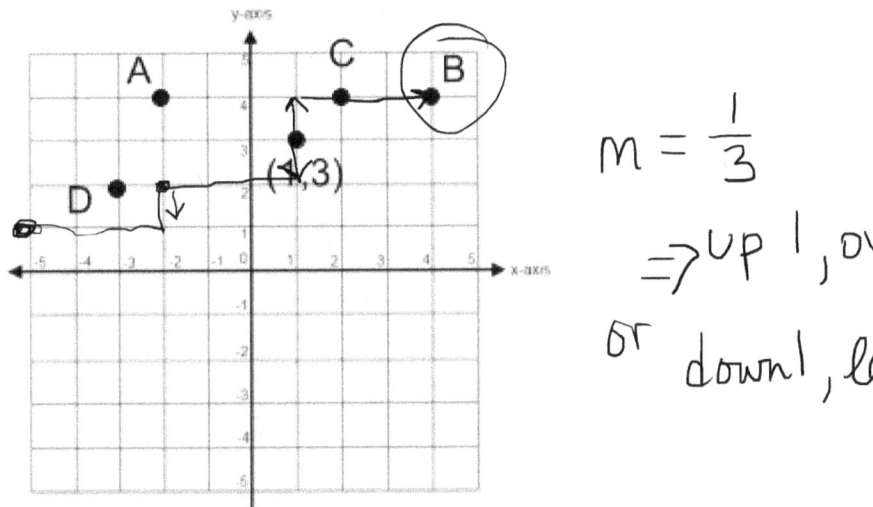

$$m = \frac{1}{3}$$

⇒ up 1, over 3

or down 1, left 3

   A. A
   (B.) B
   C. C
   D. D

92. Which of the following is the graph of the equation x + 2y = 8 in the standard (x,y) coordinate plane?

$2y = -x + 8$

$y = -\frac{1}{2}x + 4$

↑ slope   ↑ y-int

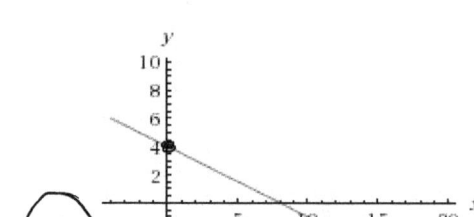
A. (circled)

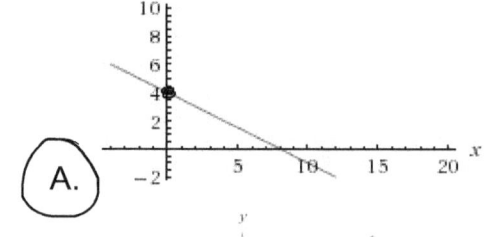
B. ✗ +slope ✗

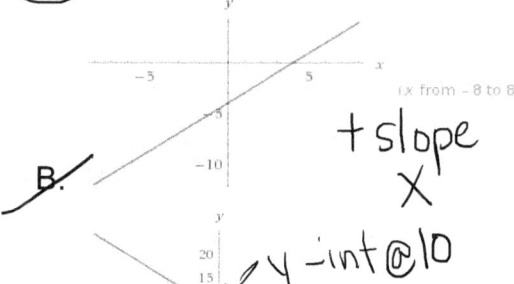
C. ✗ y-int @ 10

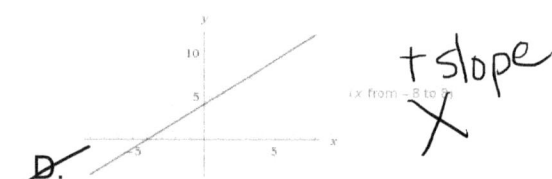
D. ✗ +slope ✗

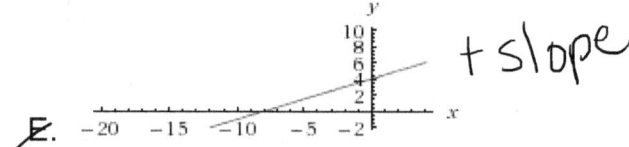
E. ✗ +slope

Inequalities

93. At a manufacturing facility, the tolerance for widget length, l, is represented by the inequality |l - 1.5| ≤ 0.01. What is the range of lengths the widget can be to pass inspection?

A. l ≥ 0.01
B. (circled) 1.49 ≤ l ≤ 1.51
C. 1.51 ≤ l
D. 1.40 ≤ l ≤ 1.60
E. 1.49 ≥ l ≥ 1.51

$|l - 1.5| \leq 0.01$

values must be within 0.01 of 1.5

$l - 1.5 \leq 0.01$
$l \leq 1.51$

and $-(l - 1.5) \leq 0.01$
$l - 1.5 \geq -0.01$
$l \geq 1.49$

94. Simplify the inequality 5(x + 1) > 2(x - 1). (On your own...Graph on a number line.)

   A. $x > -\frac{7}{3}$ ✓
   B. $x < -\frac{7}{3}$
   C. $x > \frac{7}{3}$
   D. $x > \frac{3}{7}$
   E. $x < \frac{3}{7}$

$5x + 5 > 2x - 2$
$3x > -7$
$x > -7/3$

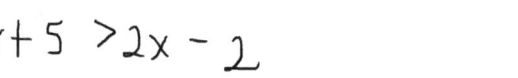

Graph of Inequality

95. Is the shaded region of the graph $y \leq x + 1$ or $y \geq x + 1$?

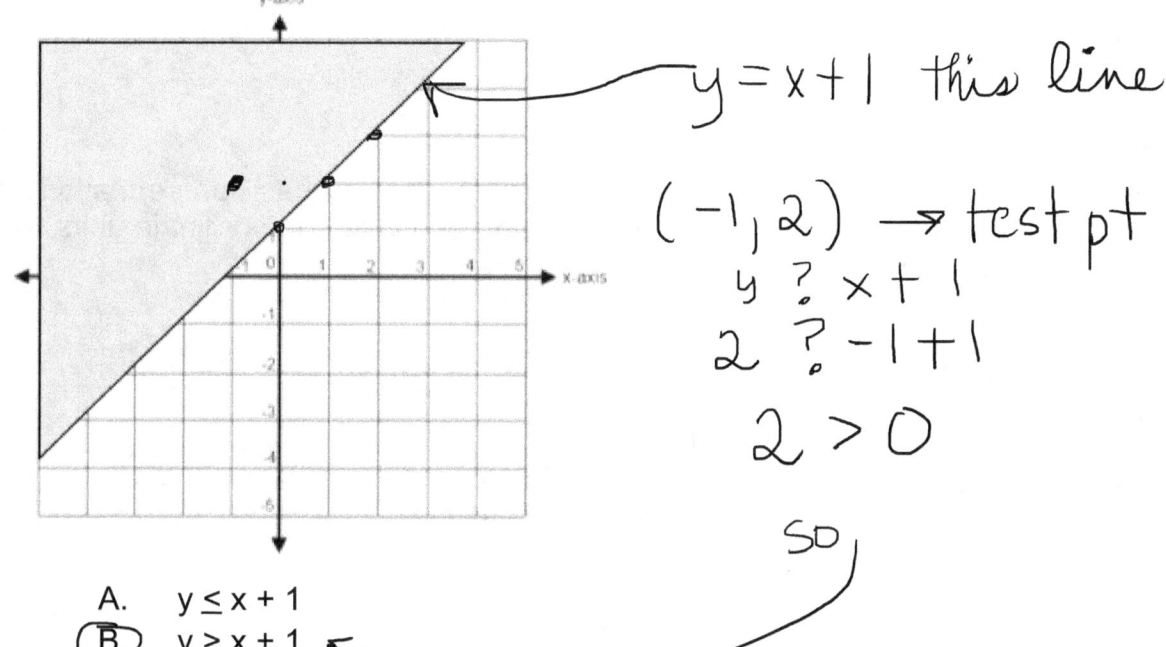

$y = x + 1$ this line

$(-1, 2) \to$ test pt
$y \; ? \; x + 1$
$2 \; ? \; -1 + 1$
$2 > 0$

SO

   A. $y \leq x + 1$
   B. $y \geq x + 1$ ✓
   C. Both
   D. Neither

96. The graphs of y = x² and y = x are shown below. What real values of x satisfy the inequality x² < x? → Where the graph $y = x^2$ is less than (below) $y = x$

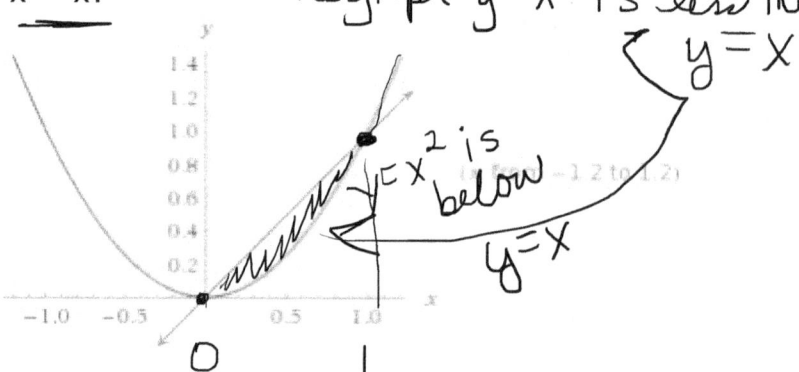

- A. x > 1
- B. x < 0
- C. x > .5
- **D. 0 < x < 1** ✓
- E. x < .5

Circles

97. A circle in the standard (x,y) coordinate plane has an equation of (x + 3)² + (y - 2)² = 9. What are the radius of the circle, in coordinate units, and the coordinates of the center of the circle?

- A. 9, (3, -2)
- B. 3, (3, -2)
- **C. 3, (-3, 2)** ✓
- D. 9, (-3, 2)
- E. 81, (-3, 2)

(h, k) center

$$(x - h)^2 + (y - k)^2 = r^2$$

$$(x + 3)^2 + (y - 2)^2 = 9$$

$(-3, 2)$

$r^2 = 9$
$r = 3$

98. What is the equation of the circle in the graph?

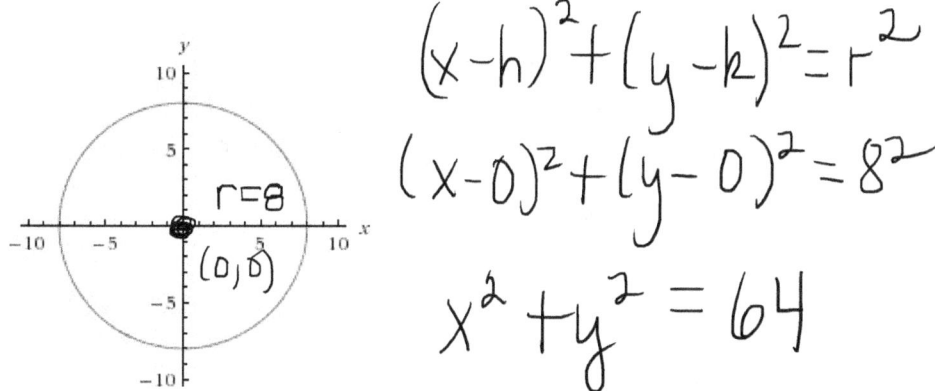

$(x-h)^2 + (y-k)^2 = r^2$

$(x-0)^2 + (y-0)^2 = 8^2$

$x^2 + y^2 = 64$

A. $(x - 2)^2 + y^2 = 8$
B. $x^2 + y^2 = 64$ ✓
C. $x^2 + y^2 = 8$
D. $(x - 5)^2 + (y - 5)^2 = 25$
E. $x^2 + y^2 = 25$

99. What is the center of the circle given by the equation $(x - 3)^2 + (y + 2)^2 = 25$?

A. (-1, 2)
B. (-3, 2)
C. (3, -2) ✓
D. (4, -1)
E. (5, -2)

$(x-h)^2 + (y-k)^2$

(h, k)

$(3, -2)$

Distance

100. What is the length of the line segment with endpoints (3,7) and (-2, 4) in the standard (x,y) coordinate plane?

A. 5
B. $\sqrt{34}$ ✓
C. 7
D. $\sqrt{53}$
E. 8

$D = \sqrt{(3+2)^2 + (7-4)^2}$

$= \sqrt{5^2 + 3^2} = \sqrt{25+9}$

$= \sqrt{34}$

Functions

Arithmetic/Geometric Sequence

101. What is the next term in the arithmetic sequence {1, -3, -7, -11, ?}?

   A. -20
   B. -17
   C. -15
   D. 5
   E. 7

102. What is the missing term in the following geometric sequence: {1, $\frac{1}{2}$, $\frac{1}{4}$, ?, $\frac{1}{16}$, $\frac{1}{32}$}?

   A. $\frac{1}{14}$
   B. $\frac{1}{12}$
   C. $\frac{1}{10}$
   D. $\frac{1}{8}$
   E. $\frac{1}{6}$

103. What is the sum of the first 5 terms of the arithmetic sequence in which the 7th term is 6.5 and the 12th term is 10.25?

   A. 3.25
   B. 5
   C. 16.75
   D. 17.5
   E. 21.25

Logs

104. What is the value of $\log_3 27$?

   A. 2
   B. 3
   C. 6
   D. 9
   E. 13

SOLUTIONS - TeachMathWithMe.com

105. What is the real value of x in the equation $\log_2 16 - \log_2 4 = \log_3 x$?

    A. 6
    B. 9 ✓
    C. 12
    D. 15
    E. 20

$\log_2 \frac{16}{4} = \log_3 x$
$\log_2 4 = \log_3 x = y$
$2^y = 4 \Rightarrow \log_3 x = 2$
$y = 2$
$3^2 = x$
$x = 9$

106. For what real value of x is $\log_{x+1}(x^2 + 5) = 2$?

    A. -2
    B. -1
    C. 0
    D. 1
    E. 2 ✓

$(x+1)^2 = x^2 + 5$
$x^2 + 2x + 1 = x^2 + 5$
$2x = 4$
$x = 2$

107. When is $f(x) = \log_b(x + 3)$ undefined?

    A. $x \leq -3$ ✓
    B. $x \geq -3$
    C. $x \leq 0$
    D. $x < -3$
    E. $x > -3$

$(x+3)$ cannot be ≤ 0
$x + 3 \leq 0$
$x \leq -3$

Reading Graphs

108. Widgets x and y are manufactured daily on a machine. The daily machine capacity is 10 ($x + y \leq 10$). If the business earns $200 for every x produced and $300 for every y produced, what is the maximum profit they can earn from x and y widgets in a day?

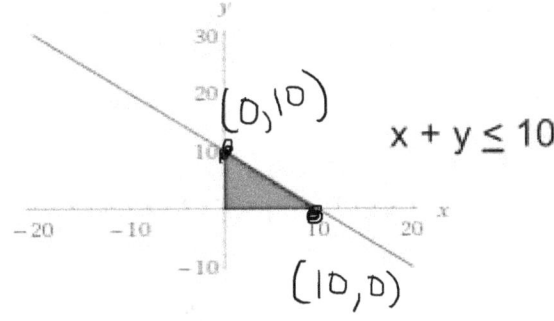

$x + y \leq 10$

(0,10)
(10,0)

    A. $300
    B. $500
    C. $2000
    D. $3000 ✓
    E. $5000

$(10, 0) \Rightarrow 200(10) + 300(0) = 2000$

$(0, 10) \Rightarrow 200(0) + 300(10) = 3000$ ✱

Solving Functions

109. If $f(x) = \sqrt{x+1}$ and $g(x) = 2x - b$ and in the standard (x,y) coordinate plane $y = f(g(x))$ passes through (5,4), what is the value of b?

A. -8
B. -5 ✓
C. 0
D. 4
E. 5

$f(g(x)) = f(2x-b)$
$= \sqrt{2x-b+1}$
$(5,4) \Rightarrow 4 = \sqrt{2(5)-b+1}$
$(4)^2 = (\sqrt{11-b})^2$
$16 = 11 - b$
$b = -5$

110. Find f(-2) if $f(x) = -x^2 + 4x$.

A. -12 ✓
B. -4
C. 4
D. 6
E. 12

$f(-2) = -(-2)^2 + 4(-2)$
$= -4 - 8 = -12$

111. Find $f(g(x))$ if $f(x) = x^2 + 3x$ and $g(x) = x + 7$.

A. $x^2 + 17x + 70$ ✓
B. $x^2 + 3x + 49$
C. $4x + 7$
D. $x^2 + 14x + 49$
E. $x + 7$

$f(x+7) = (x+7)^2 + 3(x+7)$
$= x^2 + 14x + 49 + 3x + 21$
$= x^2 + 17x + 70$

112. A function H is defined as follows:

$H(x) = x^2 + 3x - 1$, for $x > 1$ ← 2
$H(x) = -x^2 - 2x + 1$, for $x \leq 1$

What is the value of H(2)?

A. -2
B. -1
C. 1
D. 9 ✓
E. 10

$H(2) = (2)^2 + 3(2) - 1$
$= 4 + 6 - 1 = 9$

Asymptotes

113. What type of asymptote does the function f(x) = (x² + 3x +2)/(x-2) have?

    A. Vertical Only
    B. Horizontal Only
    C. Oblique Only
    D. Vertical and Horizontal
    E. Vertical and Oblique ← (circled)

$f(x) = \dfrac{x^2 + 3x + 2}{x - 2}$

V.A. at x = 2
O.A. b/c can divide num/deno (larger expo in the numerator)

114. What is/are the asymptote(s) of the function f(x) = 1/(x-2)?

    A. x = 0 only
    B. x = 2 only
    C. x = 2 and y = 0 ← (circled)
    D. x = 0 and y = 0
    E. none

$f(x) = \dfrac{1}{x-2}$

x - 2 = 0
⟹ x = 2
V.A.

+ H.A. y = 0
(larger expo on x in the deno)

Even/Odd Functions

115. Below is the graph of y=cosx. Is this function even, odd, or neither?

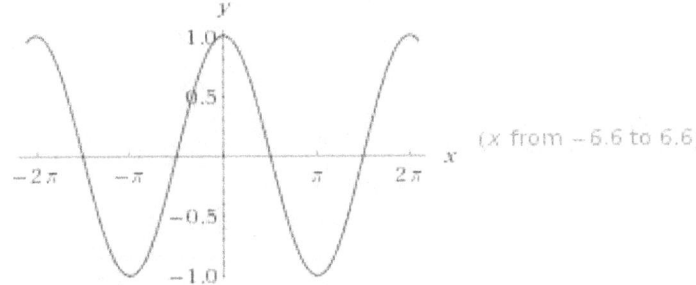
(x from –6.6 to 6.6)

    A. Even ← b/c graph is symmetric across the y-axis
    B. Odd
    C. Neither
    D. Even and Odd
    E. Unable to determine

116. A function f is an even function if and only if f(-x) = f(x) for all x in the domain of f. — Symmetric across the y-axis
    Which function graphed below is NOT an even function?

A. 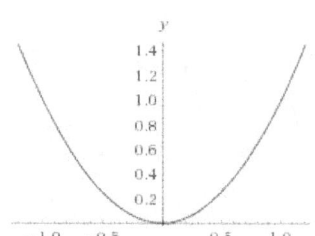 symmetric

B. 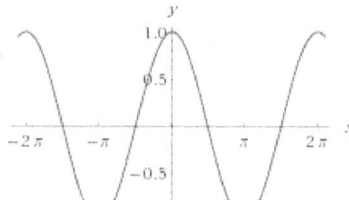 sym.

C. 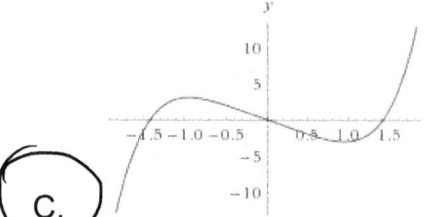 not sym. ✓ (circled)

D. 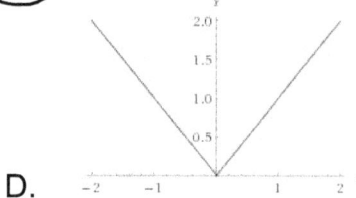  sym.

E. 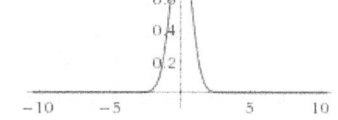 sym

117. Below is the graph of y=sinx. Is this function even, odd, or neither?

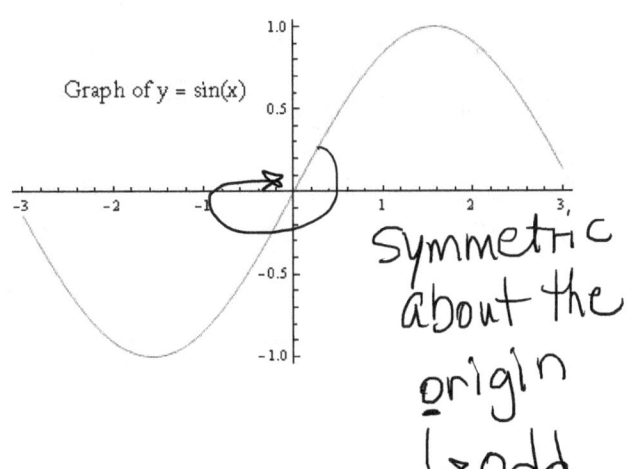

Symmetric about the origin
↳ Odd

A. Even
B. Odd
C. Even and odd
D. Neither
E. Unable to determine

(B circled)

Transformations

118. Given $y_1(t) = a_1\sin(b_1 t)$ and $y_2(t) = a_2\cos(b_2 t)$, how do b_1 and b_2 compare?

$y_1(t) = a_1\sin(b_1 t)$ $y_2(t) = a_2\cos(b_2 t)$

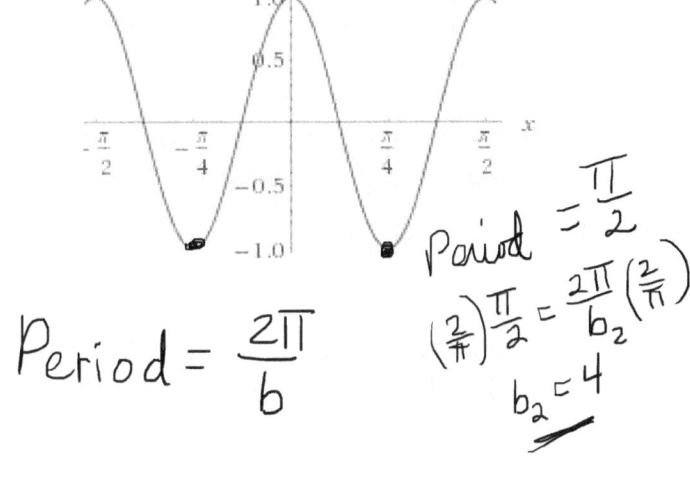

Period = 2π
$b_1 = 1$

Period = $\frac{2\pi}{b}$

Period = $\frac{\pi}{2}$

$\left(\frac{2}{\pi}\right)\frac{\pi}{2} = \frac{2\pi}{b_2}\left(\frac{2}{\pi}\right)$

$b_2 = 4$

A. $b_1 > b_2$
B. $b_2 > b_1$
C. $b_2 = b_1$
D. Unable to determine

(B circled)

119. What is an equation for a cos graph with a period of 2π, an amplitude of 2, and an upward shift of 3 units?

   A. y = 2 cos (2πx) - 3
   B. y = -2 cos (2πx) - 3
   C. y = 3 cos (x) + 3
   D. y = -2 cos (x) + 3
   E. y = -2 cos (2x) + 3

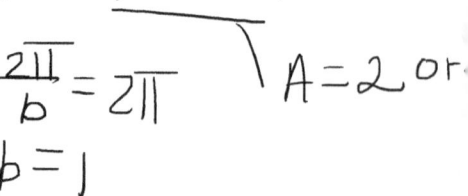

Geometry

Quadrants

120. If point C has a non-zero x-coordinate and a non-zero y-coordinate and the coordinates have the same signs, then point C must be located in which of the 4 quadrants labeled below?

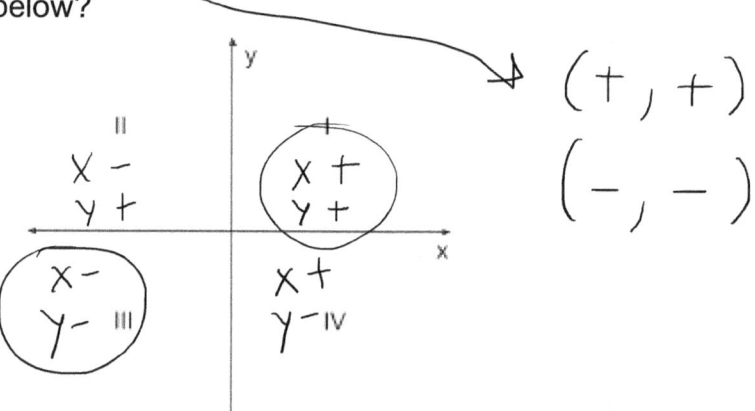

   A. I only
   B. II and III
   C. I and III
   D. I and IV
   E. III only

121. What are the quadrants of the standard (x,y) coordinate plane below that contain points on the graph of the equation x - y = 2?

   A. III only
   B. I and III only
   C. I, II, III, and IV
   D. II and IV only
   E. I, III, and IV only

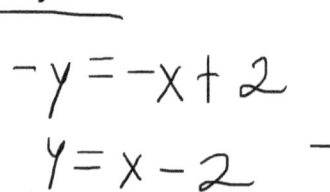
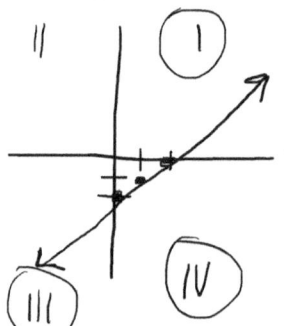

Planes

122. Two distinct planes that are not parallel intersect to form which of the following?

    A. point
    B. angle
    C. line
    D. line segment
    E. ray

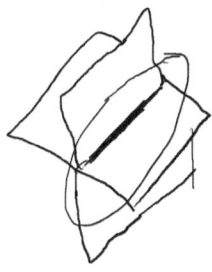

Midpoint

123. In the standard (x,y) coordinate plane, the midpoint of AB is (4,1) and A is located at (2,3). If (x,y) are the coordinates of B, what is the value of x - y?

    A. -1
    B. 0
    C. 4
    D. 6
    E. 7

$$(4,1) = \left(\frac{2+x_B}{2}, \frac{3+y_B}{2}\right)$$

$6-(-1) = 7$

$\frac{2+x}{2} = 4 \Rightarrow 2+x = 8 \Rightarrow x = 6$

$\frac{3+y}{2} = 1 \Rightarrow 3+y = 2 \Rightarrow y = -1$

124. In the standard (x,y) coordinate plane, point M with coordinates (6,1) is the midpoint of AB and B has coordinates (7,3). What are the coordinates of A?

    A. (5, -1)
    B. (1, 3)
    C. (1, -1)
    D. (6, -1)
    E. (5, 1)

$$(6,1) = \left(\frac{x_A+7}{2}, \frac{y_A+3}{2}\right)$$

$\frac{x_A+7}{2} = 6 \qquad \frac{y_A+3}{2} = 1$

$x_A + 7 = 12 \qquad y_A + 3 = 2$

$x_A = 5 \qquad y_A = -1$

$(5, -1)$

Parallel Lines

125. In the figure below, lines l and m are parallel. Line t is a transversal that intersects both l and m. Give the sets of angles with equivalent angle measures.

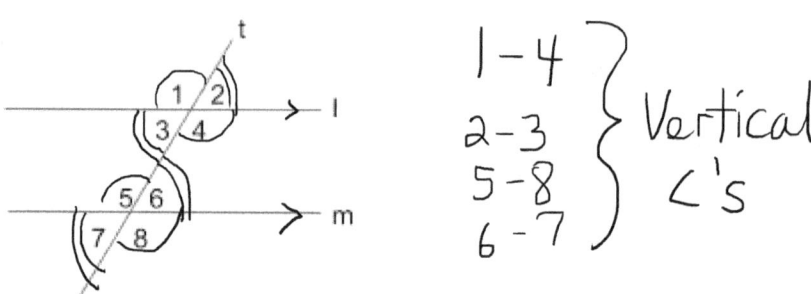

1-4
2-3
5-8
6-7 } Vertical ∠'s

A. {1, 2}, {3, 4}
B. {5, 7}, {1, 3}
C. {1, 3, 5, 7}, {2, 4, 6, 8}
(D.) {1, 4, 5, 8}, {2, 3, 6, 7}
E. {1, 2, 3, 4}, {5, 6, 7, 8}

2-6
4-8
1-5
3-7 } Corresponding ∠'s

126. In the figure below, lines l and m are parallel and intersected by lines s and t. Given the angle measures below, what is the measure of angle x?

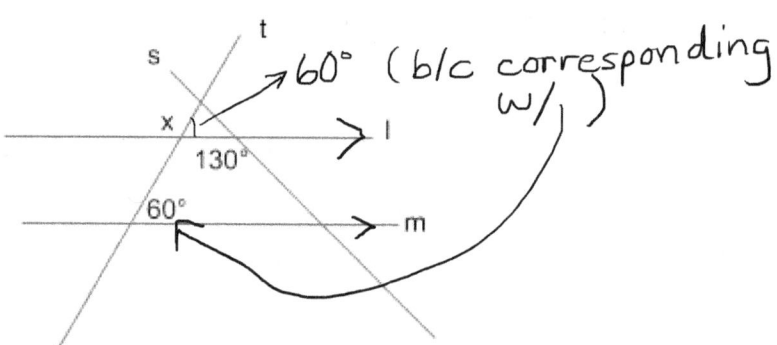

60° (b/c corresponding w/)

A. 30°
B. 60°
C. 90°
(D.) 120°
E. 130°

X = 180 - 60 = 120°

*supplementary relationship

N-Sided Polygons

127. If 2 angles of a pentagon measure 70° and 100°, what is the sum of the other angles of the pentagon?

   A. 70°
   B. 100°
   C. 170°
   D. 370° ← (circled)
   E. 540°

$(n-2)180° = (5-2)180 = 540°$ (sum of ∠'s of a pentagon)

$540 - 70 - 100 = 370°$

Surface Area

128. If the surface area of a square cube is 96 in², how long, in inches, are the edges of the cube?

   A. 4 ← (circled)
   B. 9.8
   C. 16
   D. 24
   E. 48

6 sides → $\frac{96}{6} = 16$ in²/face

$x^2 = 16$
$x = 4$

129. What is the total surface area of this right circular cylinder, in square inches? (Note: The total surface area of a cylinder is given by $2\pi r^2 + 2\pi rh$ where r is the radius and h is the height.)

r = 10, 30 = h, 20

   A. 650π
   B. 800π ← (circled)
   C. 800π²
   D. 850π
   E. 850π²

$A = 2\pi r^2 + 2\pi rh$
$= 2\pi(10)^2 + 2\pi(10)(30)$
$= 200\pi + 600\pi$
$= 800\pi$

Angles

130. In the isosceles trapezoid ABCD, AB is parallel to DC, ∠BDC measures 35°, and ∠BCA measures 25°. What is the measure of ∠DBC?

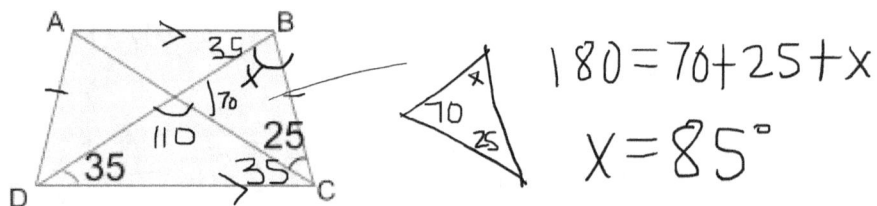

A. 25°
B. 35°
C. 85°
D. 120°
E. Cannot be determined

131. Given the angle measures below, if points O, P, R, and S are collinear, what is the measure of ∠QRS?

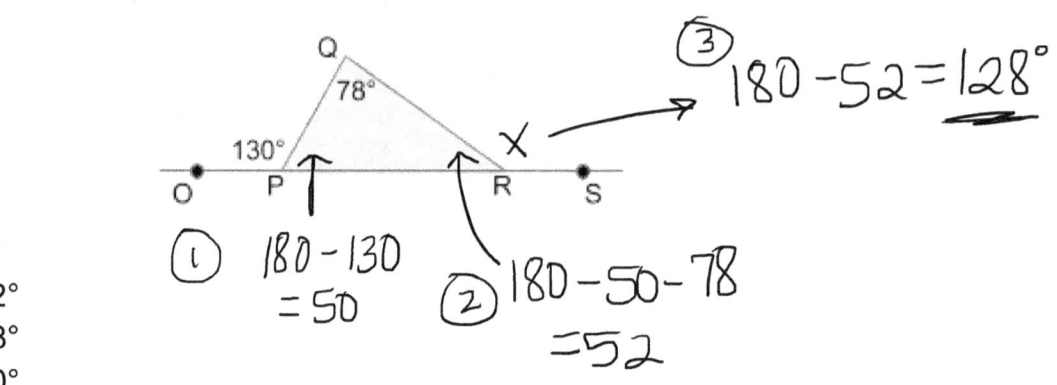

A. 78°
B. 102°
C. 128°
D. 130°
E. Cannot be determined

132. For all △'s ABC where ∠A < ∠B, how do the lengths of BC and AC compare?

A. BC ≥ AC
B. BC = AC
C. BC > AC
D. BC < AC
E. Cannot be determined

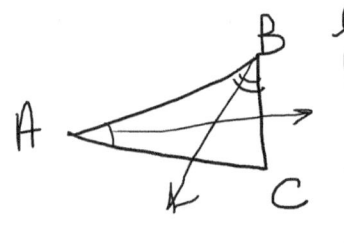

Pythagorean Theorem

133. In the right triangle pictured below, what is the length of the hypotenuse?

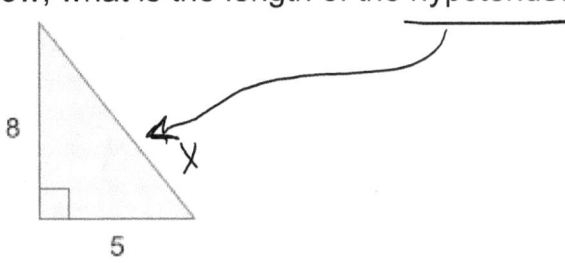

A. $\sqrt{13}$
B. $\sqrt{89}$ ← (circled)
C. 13
D. 54
E. 89

$8^2 + 5^2 = x^2$
$64 + 25 = x^2$
$\sqrt{x^2} = \sqrt{89}$
$x = \sqrt{89}$

134. Given an isosceles right triangle with one leg = 5 units, what is the length of the hypotenuse?

A. 10
B. 5
C. $5\sqrt{2}$ ← (circled)
D. 10
E. Unable to determine

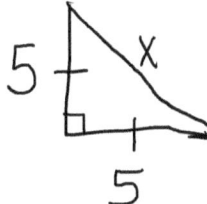

$5^2 + 5^2 = x^2$
$25 + 25 = x^2$
$x^2 = 50$
$x = \sqrt{50} = 5\sqrt{2}$

135. What is the length of a leg of a right triangle with hypotenuse of 13 and one leg 11?

A. 1
B. $2\sqrt{3}$
C. 3.2
D. $4\sqrt{3}$ ← (circled)
E. 5

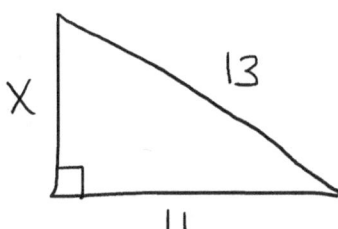

$x^2 + 11^2 = 13^2$
$\sqrt{x^2} = \sqrt{13^2 - 11^2} = \sqrt{48}$
$x = 4\sqrt{3}$

136. What is an expression for *a* in terms of *b* and *c*?

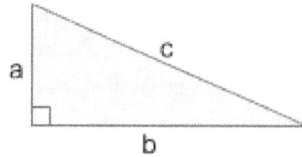

A. $\sqrt{b^2 + c^2}$
B. $\sqrt{(-b^2 + a^2)}$
C. $\sqrt{(-a^2 + c^2)}$
D. $\sqrt{(-c^2 + b^2)}$
E. $\sqrt{(-b^2 + c^2)}$ ← (circled)

$$a^2 + b^2 = c^2$$
$$\sqrt{a^2} = \sqrt{-b^2 + c^2}$$
$$a = \sqrt{-b^2 + c^2}$$

Similar Triangles

137. What is the perimeter of △PQR? △ABC ~ △PQR (Note: The symbol ~ means "is similar to".)

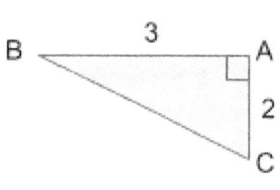

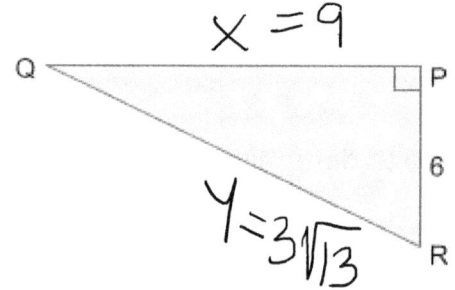

$x = 9$

$y = 3\sqrt{13}$

A. $5 + \sqrt{13}$
B. 9
C. $15 + \sqrt{13}$
D. $15 + 3\sqrt{13}$ ← (circled)
E. 32

$$\frac{2}{6} = \frac{3}{x}$$
$$2x = 18$$
$$x = 9$$

$$y^2 = 9^2 + 6^2$$
$$y^2 = 81 + 36$$
$$y^2 = 117$$
$$y = \sqrt{117}$$
$$y = 3\sqrt{13}$$

$$P = 9 + 6 + 3\sqrt{13}$$
$$= 15 + 3\sqrt{13}$$

138. In right triangle PQR, ST is parallel to PQ, and ST is perpendicular to QR at T, PR is 30 inches, ST is 3 inches, and TR is 4 inches. What is the length, in inches, of PQ?

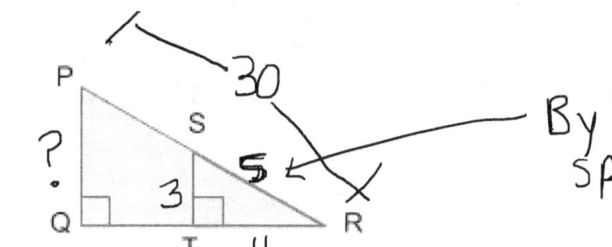

By 3-4-5 special right triangle

A. 6
B. 12
C. 18
D. 25
E. 30

$$\frac{3}{x} = \frac{5}{30}$$

$$5x = 90$$

$$x = 18$$

Volume

139. Julie is packing items that she has sold online. She needs to find a box that her product will fit inside. One box she is looking at holds a volume of 4032 cubic inches. The box description says it has a length of 18" and a width of 16". How tall, in inches, is this box?

A. 12"
B. 14"
C. 16"
D. 18"
E. 20"

$$V = L \times W \times H$$

$$4032 = 18(16)(H)$$

$$H = 14"$$

140. The volume of a right cylinder is 37.68 cubic feet. If the height of the cylinder is 3 feet, what is the diameter, in feet? Use 3.14 for π.

A. 2
B. 3.7
C. 4
D. 6
E. 15.2

$$V = \pi r^2 h$$

$$37.68 = (3.14)(r^2)(3)$$

$$r^2 = 4$$

$$r = 2$$

$$d = 2(2) = 4$$

Circles

141. As shown in the standard (x,y) coordinate plane, P(3,7) lies on the circle with center (3,3) and radius 4 coordinate units. What are the coordinates of P after the circle is rotated 180° about the center of the circle?

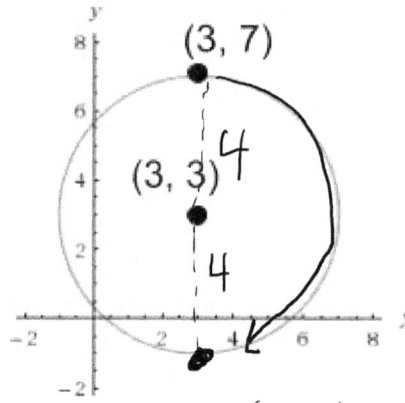

$(3,3) - 4$ y units
$\Rightarrow (3, 3-4) = (3, -1)$

A. (3, -.5)
B. (3, -1) ✓
C. (7, 3)
D. (-1, 3)
E. (5.5, 0)

Circle - Central Angles and Arcs

142. What is the measure of the central angle x?

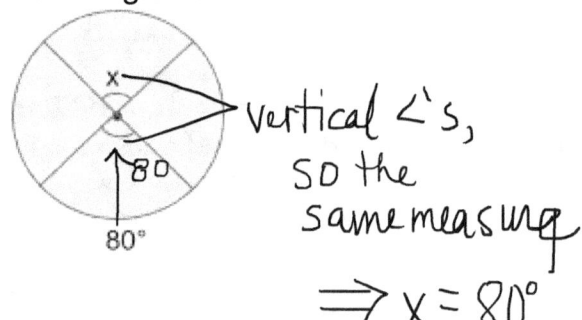

vertical ∠'s, so the same measure
$\Rightarrow x = 80°$

A. 20°
B. 40°
C. 80° ✓
D. 160°
E. Unable to determine

SOLUTIONS - TeachMathWithMe.com

143. Given a standard wall clock, as in the figure below, what is the arc measure, in degrees, of the arc between 3 and 5?

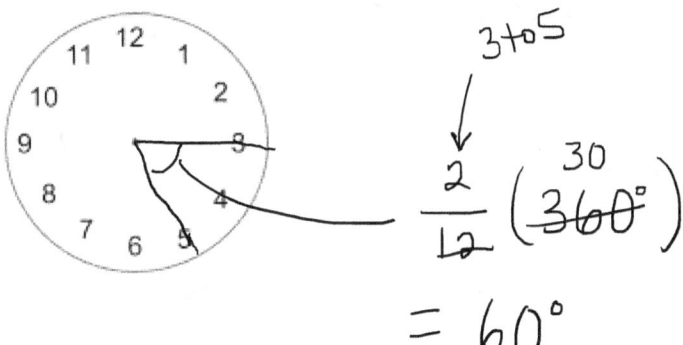

A. 15
B. 20
C. 30
D. 45
E. 60 ✓

$$\frac{2}{12}(360°) = 60°$$

144. Given the figure below, what is the measure of the cut out arc length, in inches, to the nearest tenth?

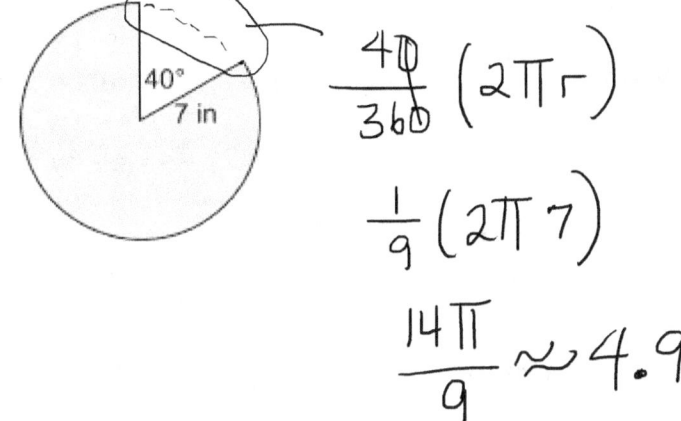

A. 4.9 ✓
B. 5.7
C. 21.3
D. 43.9
E. 44.2

$$\frac{40}{360}(2\pi r)$$
$$\frac{1}{9}(2\pi \cdot 7)$$
$$\frac{14\pi}{9} \approx 4.9$$

Inscribed Angles

145. In the figure shown below, what is the degree measure of arc QR?

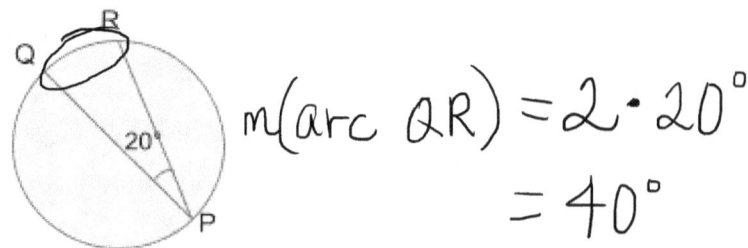

m(arc QR) = 2 · 20°
= 40°

- A. 10°
- B. 20°
- C. 35°
- (D.) 40°
- E. 42°

Conic Sections

146. What conic section is the following equation classified as?

$$x^2 + 2x = -3y^2 - 4y + 36$$

- A. parabola — X both x+y are squared
- B. circle
- (C.) ellipse
- D. hyperbola — X has a "+" between x^2 & y^2 — "−" for hyperbola
- E. none of the above

$x^2 + 2x + 3y^2 + 4y = 36$

$x^2 + 2x + 1 + 3(y^2 + \frac{4}{3}y\ ) = 36 + 1$

$(x+1)^2 + 3(y + \cdots$

could go further & complete the square if not sure, but looks like will have denominator on $x^2 + y^2$ as w/ an ellipse

Sine

147. Find sin θ.

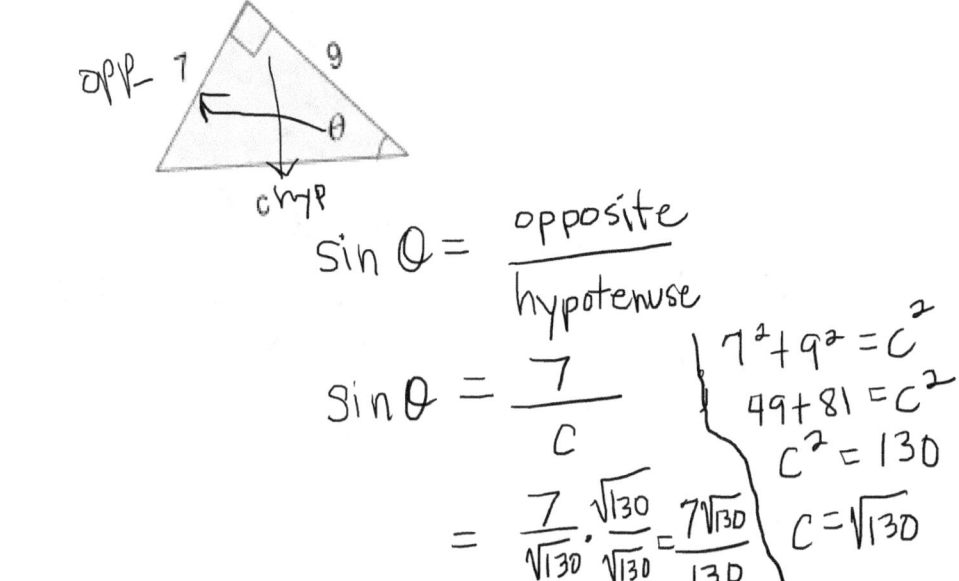

A. $\frac{c}{9}$

B. $\frac{9}{7}$

C. $\frac{7}{c}$

D. $\frac{7\sqrt{130}}{130}$ ✓

E. $\frac{\sqrt{130}}{7}$

148. Given AB and AC, what trigonometric expression gives the measure of ∠ABC?

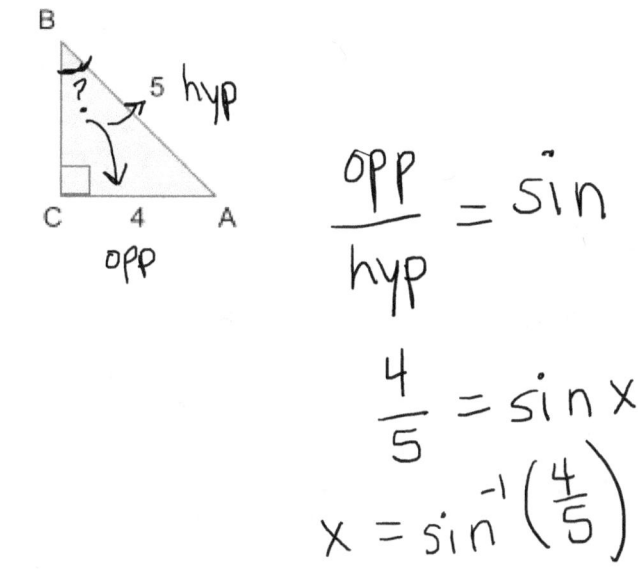

A. $\sin^{-1}\left(\frac{5}{4}\right)$

B. $\sin^{-1}\left(\frac{4}{5}\right)$ ✓

C. $\sin^{-1}\left(\frac{3}{5}\right)$

D. $\cos^{-1}\left(\frac{5}{4}\right)$

E. $\cos^{-1}\left(\frac{4}{5}\right)$

149. If $\sin \theta = 0.8$, what is the length of BD (x), to the nearest tenth?

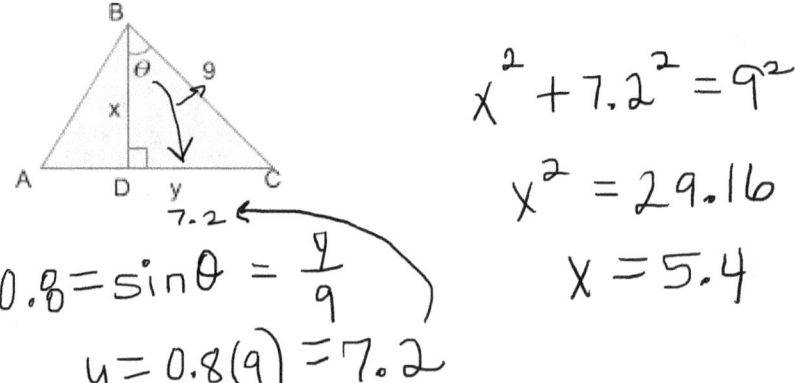

A. 1.3
B. 1.5
C. 3.2
D. 5.4
E. 7.2

$0.8 = \sin \theta = \frac{y}{9}$

$y = 0.8(9) = 7.2$

$x^2 + 7.2^2 = 9^2$

$x^2 = 29.16$

$x = 5.4$

Cosine

150. Find $\cos \theta$.

A. $\frac{2\sqrt{13}}{4}$
B. $\frac{2\sqrt{13}}{13}$
C. 4
D. $8\sqrt{13}$
E. $\frac{13\sqrt{2}}{4}$

$\cos = \frac{adj}{hyp}$

$\cos \theta = \frac{4}{2\sqrt{13}} \cdot \frac{\sqrt{13}}{\sqrt{13}} = \frac{2\sqrt{13}}{13}$

151. What expression gives the measure of $\angle B$?

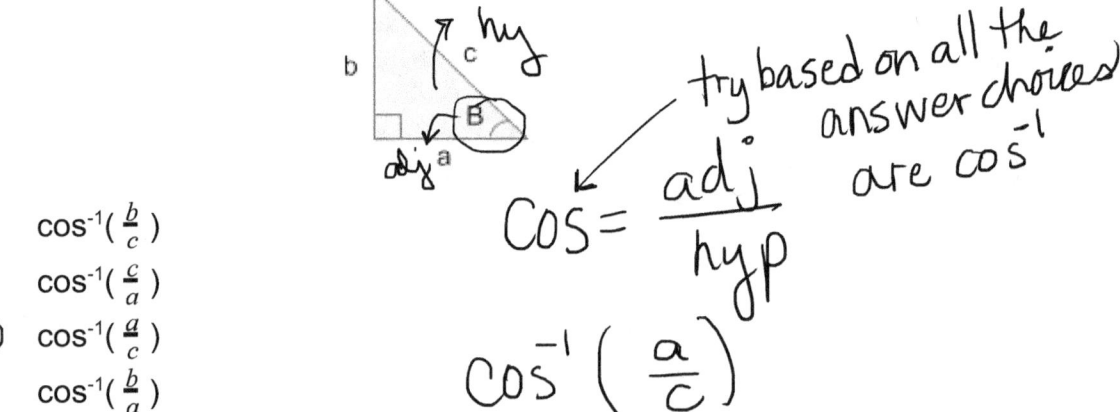

A. $\cos^{-1}(\frac{b}{c})$
B. $\cos^{-1}(\frac{c}{a})$
C. $\cos^{-1}(\frac{a}{c})$
D. $\cos^{-1}(\frac{b}{a})$
E. $\cos^{-1}(\frac{c}{a})$

try based on all the answer choices are $\cos^{-1}$

$\cos = \frac{adj}{hyp}$

$\cos^{-1}(\frac{a}{c})$

Tangent

152. Find tan A.

   A. $\dfrac{7}{3}$

   B. $\dfrac{3}{7}$

   C. $\dfrac{3\sqrt{10}}{20}$ ✓

   D. $\dfrac{2\sqrt{10}}{3}$

   E. $\dfrac{2\sqrt{10}}{7}$

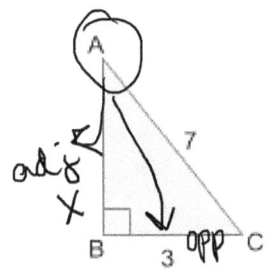

$\tan = \dfrac{opp}{adj}$

$x^2 + 3^2 = 7^2$

$x^2 = 40$

$x = 2\sqrt{10}$

$\tan A = \dfrac{3}{2\sqrt{10}} \cdot \dfrac{\sqrt{10}}{\sqrt{10}} = \dfrac{3\sqrt{10}}{20}$

153. One of the angle measures is $\tan^{-1}(\tfrac{b}{a})$. What is $\sin(\tan^{-1}(\tfrac{b}{a}))$?

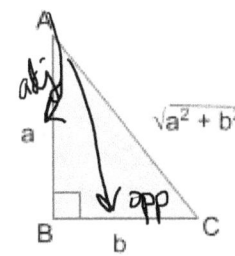

$\tan = \dfrac{opp}{adj}$ $\dfrac{b}{a} \Rightarrow \angle A$

$\tan^{-1}(\tfrac{b}{a}) = A$

$\sin A = \dfrac{b}{\sqrt{a^2+b^2}}$

   A. $\dfrac{b}{\sqrt{a^2+b^2}}$ ✓

   B. $\dfrac{b}{a}$

   C. $\sqrt{a^2+b^2}$

   D. $\dfrac{a}{b}$

   E. Unable to determine

Trigonometric Identities

154. Simplify the expression $(\sin x + \cos x)^2 - 1$ using a Pythagorean Identity.

   A. 0
   B. $\sin^2 x$
   C. 2sinx cosx ✓
   D. 2 cosx
   E. 2 sinx

$(\sin x + \cos x)(\sin x + \cos x) - 1$

$\sin^2 x + 2\sin x \cos x + \cos^2 x - 1$

$\underbrace{\sin^2 x + \cos^2 x}_{1} + 2\sin x \cos x - 1$

$1 + 2\sin x \cos x - 1$

$2\sin x \cos x$

155. Rewrite the expression in terms of 1 function. $\frac{\sin^2 x}{\csc^3 x}$

   A. sinx
   B. sin²x
   C. sin⁵x ← (circled)
   D. csc⁵x
   E. Unable to determine

$$\frac{\sin^2 x}{\frac{1}{\sin^3 x}} = \sin^2 x \cdot \frac{\sin^3 x}{1}$$

$$= \sin^{(2+3)} x = \sin^5 x$$

Law of Sines

156. In △PQR, what is the expression for the length of QR? (Note: The law of sines states that, for any triangle, the ratios of the sines of the interior angles to the lengths of the sides opposite those angles are equal.)

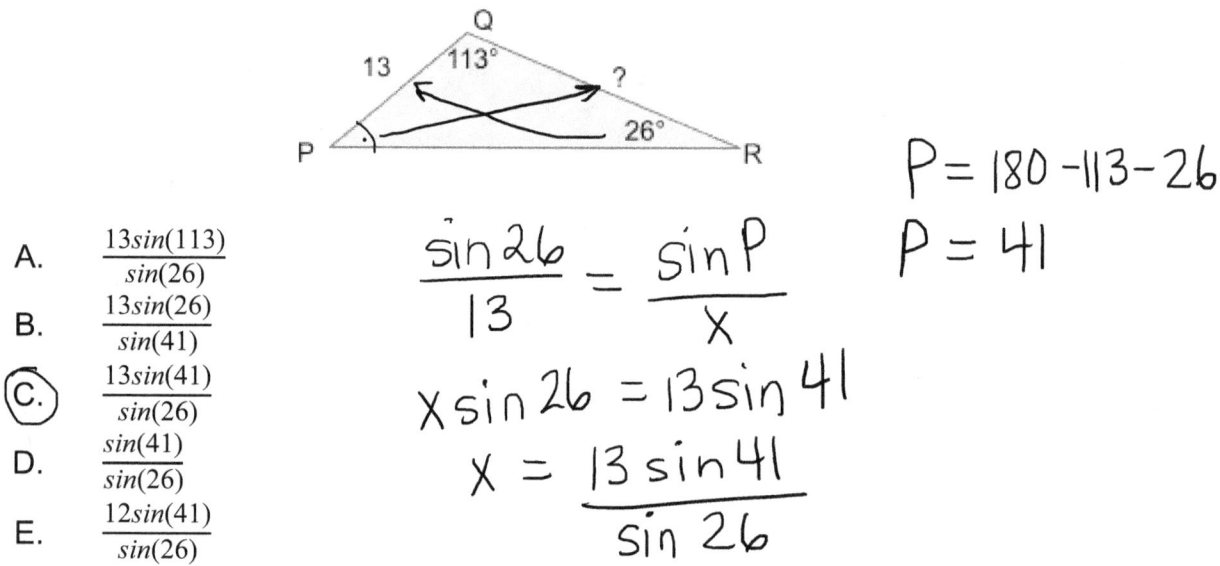

   A. $\frac{13\sin(113)}{\sin(26)}$
   B. $\frac{13\sin(26)}{\sin(41)}$
   C. $\frac{13\sin(41)}{\sin(26)}$ ← (circled)
   D. $\frac{\sin(41)}{\sin(26)}$
   E. $\frac{12\sin(41)}{\sin(26)}$

$\frac{\sin 26}{13} = \frac{\sin P}{x}$

$P = 180 - 113 - 26$
$P = 41$

$x \sin 26 = 13 \sin 41$

$x = \frac{13 \sin 41}{\sin 26}$

157. In △ABC, AB is 20 units long, AC is 13 units long, and the measure of ∠B is 22°. What is the measure of ∠A, to the nearest tenth?

   A. 2.0°
   B. 35.19°
   C. 55.0°
   D. 122.8° ← (circled)
   E. 125.3°

*No indication that this is a right triangle.

② ∠A = 180 − 22 − 35.2 = 122.8°   *2 unknowns

$\left\{ \frac{\sin A}{x} = \frac{\sin 22}{13} \right.$ ⟹ ① $\frac{\sin 22}{13} = \frac{\sin C}{20}$

$13 \sin C = 20 \sin 22$

$\sin C = \frac{20 \sin 22}{13}$

$C = \sin^{-1}\left(\frac{20 \sin 22}{13}\right)$

$C \approx 35.2$

SOLUTIONS - TeachMathWithMe.com

Law of Cosines

158. What is the measure of ∠A, to the nearest tenth?

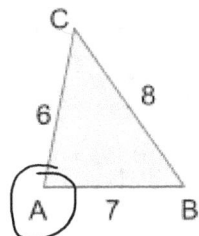

*No indication of a right triangle

A. 15.2
B. 63.3
C. 75.5 ✓
D. 83.3
E. 122.1

$8^2 = 6^2 + 7^2 - 2(6)(7)\cos A$

$A = \cos^{-1}\left(\dfrac{8^2 - 6^2 - 7^2}{-2(6)(7)}\right) \approx 75.5°$

159. If a = 6, c = 9, and B = 128°, then what is b, to the nearest tenth?

A. 10
B. 11.3
C. 13.5 ✓
D. 20
E. 183.5

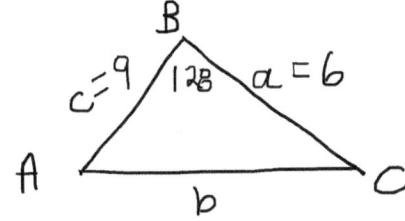

$b^2 = 9^2 + 6^2 - 2(9)(6)\cos 128$

$b \approx 13.5$

Statistics and Probability

Fundamental Counting Principle

160. In a florist, customers can choose from 5 types of roses, 3 types of greenery, and 4 colors of tissue paper. How many different bouquets are possible?

A. 5
B. 12
C. 60 ✓
D. 65
A. 100

$(5)(3)(4) = 60$

161. There are 5 students in a class. How many different ways can the 5 students form a line?

A. 5
B. 6
C. 10
D. 15
E. 120 ✓

$5! = 5(4)(3)(2)(1) = 120$

162. A customer must set a 4-digit PIN for his bank account using the number 0-9. How many possible outcomes are there for the PIN?   0-9 → 10 choices

A. 9^4
B. 10^4 ✓
C. 10^{10}
D. 10^{14}
E. 10^{40}

Combinations

$\underline{0\text{-}9}\ \underline{0\text{-}9}\ \underline{0\text{-}9}\ \underline{0\text{-}9}$

$\underline{10}\ \underline{10}\ \underline{10}\ \underline{10}$

10^4

163. A committee will be selected from a group of 15 women and 12 men. The committee will be made up of 2 women and 2 men. Which of the following expressions gives the number of different committees that could be selected from these 27 people?

A. $_{27}C_4$
B. $(_{15}C_2)(_{12}C_2)$ ✓
C. $(_{27}C_2)(_{27}C_2)$
D. $(_{15}P_2)(_{12}P_2)$
E. $_{27}P_4$

Permutations

order does NOT matter → Combination

Women   men
$(_{15}C_2)\ (_{12}C_2)$

15 pick 2, 12 pick 2

164. How many ways can we award a 1st place ribbon and a 2nd place ribbon among 20 contestants?

A. $_{10}C_2$
B. $_{20}C_2$
C. $_{10}P_2$
D. $_{18}P_2$
E. $_{20}P_2$ ✓

Probability

order of choosing 1st + 2nd does matter ⇒ Permutation

$_{20}P_2$

20 pick 2

165. An integer from 1-100 is to be chosen randomly. What is the probability, to the nearest hundreth, that the number chosen will have 1 as at least 1 digit?

A. 0.15
B. 0.20 ✓
C. 0.22
D. 0.33
E. 0.50

1-100 — 100 total numbers (sample size)

possibilities: 1, 10, 11, 12, 13, 14, 15, 16, 17, 18, 19, 21, 31, 41, 51, 61, 71, 81, 91, 100 } 20

Prob = $\frac{20}{100}$ = .20

166. A security code is generated with the format of 1 numerical digit (0-9), 5 alphabetical letters, and 1 numerical digit (0-9). The numerical digits can repeat, but the alphabetical letters cannot repeat. What is the probability of generating the code with the letters "MUSIC"?

- A. $\dfrac{10^2}{(10^2)(26)(25)(24)(23)(22)}$ ⟵ (circled)
- B. $\dfrac{1}{10^2}$
- C. $\dfrac{1}{26^2}$
- D. $\dfrac{10^2}{(10^2)(26)}$
- E. $\dfrac{1}{(10^2)(26)(25)(24)(23)(22)}$

0-9 M U S I C 0-9
10   26 25 24 23 22 10

$\dfrac{10}{10}$ $\dfrac{1}{26}$ $\dfrac{1}{25}$ $\dfrac{1}{24}$ $\dfrac{1}{23}$ $\dfrac{1}{22}$ $\dfrac{10}{10}$ ← doesn't matter what digits are

Can be all 10 digits

$\dfrac{10(10)}{10^2(26)(25)(24)(23)(22)}$

Standard Deviation

167. If Matt's teacher decides to only record 4 grades for the grading period, what is the standard deviation of Matt's 4 test scores: 92, 85, 96, and 100?

- A. 1.39
- B. 5.54 (circled)
- C. 11.08
- D. 93.25
- E. 122.75

① Calculate w/ calculator
or
② By hand
avg = $\dfrac{373}{4}$ = 93.25

$\sqrt{\dfrac{(92-93.25)^2 + \cdots (100-93.25)^2}{4}}$

168. Sarah has 12 rose bushes, but decides to record the number of blooms from only 4 of these plants over a certain time period. When she has all the data she plans to collect, she decides to calculate the standard deviation of the number of blooms per rose bush. What number will she divide the sum of squared differences by?

- A. 3 (circled)
- B. 4
- C. 12
- D. 20
- E. 48

4 is her sample size: (n)

For a sample, divide by n-1 → 4-1 = 3

169. Which set of numbers has a smaller standard deviation? *Try to look for a smaller spread

- A. 8, 12, 14, 3, 1, 6, 9   1-14
- B. 5, 1, 7, 9, 4, 12, 2   1-12
- C. 10, 5, 1, 12, 11, 2, 1   1-12
- D. 4, 12, 8, 9, 1, 10, 3   1-12
- E. 3, 4, 6, 2, 3, 5, 4   2-6 (circled)

*Could calculate or notice that compared w/ choices A-D, E's values are all fairly close together.

Bivariate Data

170. A softball concessions director wants to determine if temperature affects snow cone sales to better gauge how many supplies to order for summer tournaments. If the statistical software she uses calculates a correlation coefficient (r) of 0.97, which scatter plot below could represent the data she collected?

(A.)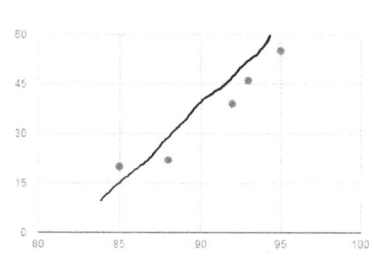

B.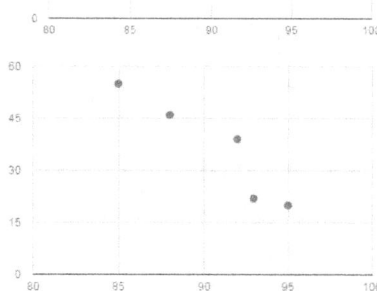

C.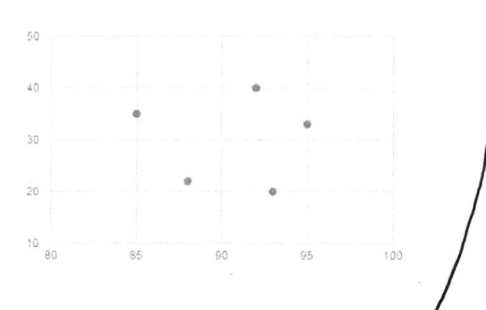

$r = 0.97$

(highly correlated — the closer $|r|$ is to 1.0)

- Since r is positive, the line of best fit is in a positive direction as in A.

171. Which scatter plot could represent data with r = 0?

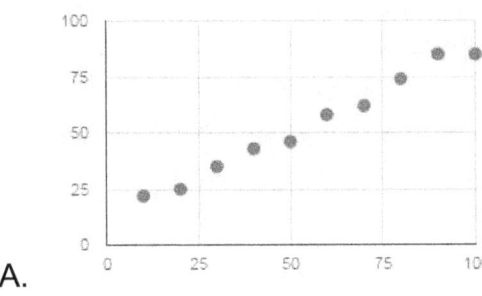

A. $r \approx 1.0$

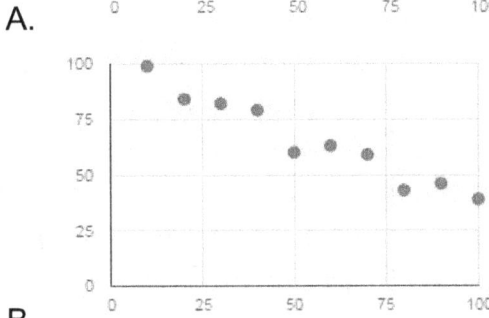

B. $r \approx -1.0$

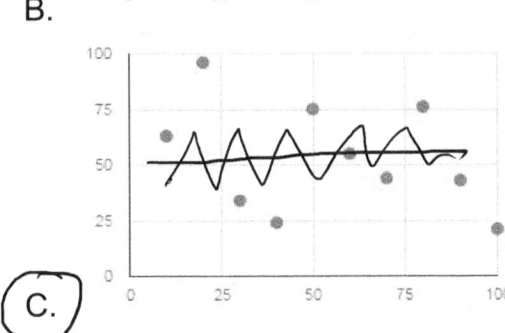

(C.)   Data is scattered & r = 0 indicates no correlation.

Answers

#	Ans	#	Ans	#	Ans	#	Ans	#	Ans	#	Ans	#	Ans
1	A	26	B	51	D	76	D	101	C	126	D	151	C
2	C	27	D	52	D	77	D	102	D	127	D	152	C
3	A	28	C	53	A	78	C	103	D	128	A	153	A
4	B	29	C	54	B	79	A	104	B	129	B	154	C
5	D	30	D	55	B	80	B	105	B	130	C	155	C
6	D	31	D	56	D	81	C	106	E	131	C	156	C
7	E	32	D	57	C	82	D	107	A	132	D	157	D
8	A	33	C	58	D	83	B	108	D	133	B	158	C
9	D	34	C	59	C	84	C	109	B	134	C	159	C
10	C	35	D	60	B	85	B	110	A	135	D	160	C
11	B	36	C	61	D	86	B	111	A	136	E	161	E
12	C	37	D	62	A	87	C	112	D	137	D	162	B
13	C	38	C	63	C	88	C	113	E	138	C	163	B
14	D	39	B	64	A	89	B	114	C	139	B	164	E
15	C	40	B	65	C	90	D	115	A	140	C	165	B
16	B	41	C	66	C	91	B	116	C	141	B	166	A
17	D	42	C	67	C	92	A	117	B	142	C	167	B
18	B	43	D	68	C	93	B	118	B	143	E	168	A
19	E	44	D	69	A	94	A	119	D	144	A	169	E
20	B	45	A	70	B	95	B	120	C	145	D	170	A
21	A	46	D	71	B	96	D	121	E	146	C	171	C
22	C	47	D	72	E	97	C	122	C	147	D		
23	C	48	B	73	C	98	B	123	E	148	B		
24	E	49	A	74	B	99	C	124	A	149	D		
25	C	50	D	75	D	100	B	125	D	150	B		

Formulas

Use this formula sheet as a guide in creating your own. Add any as needed!

Perimeter:
- Rectangle: $P = 2l + 2w$
- Square: $P = 4s$
- Triangle: $P = a + b + c$, a, b, and c are △ sides

Circumference:
- $C = \Pi d = 2\Pi r$

Area:
- Rectangle: $A = lw$
- Triangle: $A = \frac{1}{2}bh$
- Circle: $A = \Pi r^2$
- Trapezoid: $A = \frac{1}{2}h(b_1 + b_2)$

Surface Area:
- Right Cylinder: $SA = 2\Pi r^2 + 2\Pi rh$

Volume:
- Rectangle: $V = lwh$
- Cylinder: $V = \Pi r^2 h$

Log:
$$\log_a x = b \rightarrow a^b = x$$

Equation of Circle:
$(x - h)^2 + (y - k)^2 = r^2$, center: (h, k) and radius = r

Midpoint:
$$\left(\frac{x_1 + x_2}{2}, \frac{y_1 + y_2}{2} \right)$$

Slope:
$$m = \frac{y_2 - y_1}{x_2 - x_1}$$

Distance:
$$d = \sqrt{(x_2 - x_1)^2 + (y_2 - y_1)^2}$$

Pythagorean Theorem:
$$a^2 + b^2 = c^2$$

Trigonometry:
$$\sin x = \frac{opp}{hyp} \quad \cos x = \frac{adj}{hyp} \quad \tan x = \frac{opp}{adj}$$

Trig Identities:
$$\csc x = \frac{1}{\sin x} \quad \sec x = \frac{1}{\cos x} \quad \cot x = \frac{1}{\tan x}$$

$$\sin^2 x + \cos^2 x = 1$$
$$\sec^2 x = 1 + \tan^2 x$$
$$\csc^2 x = 1 + \cot^2 x$$

Law of Cosines (usually given on the test, but just in case...):
$$c^2 = a^2 + b^2 - 2ab(\cos C)$$

Law of Sines (usually given in words, but just in case...):
$$\frac{a}{\sin A} = \frac{b}{\sin B} = \frac{c}{\sin C} \quad \text{OR}$$
$$\frac{\sin A}{a} = \frac{\sin B}{b} = \frac{\sin C}{c}$$

Functions:
- Even: $f(-x) = f(x)$ → graph is symmetric about the y-axis
- Odd: $-f(x) = f(-x)$ → graph is symmetric about the origin (180° test)

Transformations:
$f(x) = A \sin(Bx - C) + D$ OR
$f(x) = A \cos(Bx - C) + D$

Amplitude = |A|
Period = $\frac{2\Pi}{B}$
Vertical Shift = D (up if positive, down if negative)
Horizontal Shift = C/B (if "C/B" is positive, move to the right; if "C/B" is negative, move to

Total Measure of Interior ∠ of N-Sided Polygon: $(n-2)\,180°$	the left; i.e., $f(x) = A\sin(Bx - C) + D$ will be shifted C/B units to the right

Final Words

Congratulations! You have made it through this practice workbook. I know you are well on your way to doing your best on the math section of the ACT (and hopefully on the other sections too!). Do not forget to keep on practicing. Here are a few other things to do in preparation for the test:

- Work more math problems.
- Make sure that you fully understand them.
- Try new techniques to help you work the problems just a little faster.

I hope this book has been helpful to you. I enjoyed putting it together, and I still want to continue to make it even better. I would love to hear what you think about it!

Email me at contact@teachmathwithme.com.

Thank you again, and I wish you the best on your ACT journey!

Mandee Boster

P.S. Be sure to check out the videos on my website if you want to see me work some of the questions in the workbook!

www.ingramcontent.com/pod-product-compliance
Lightning Source LLC
Chambersburg PA
CBHW081431220526
45466CB00008B/2343